Advanced Series in Agricultural Sciences 3

Raoul A. Robinson

Plant Pathosystems

With 15 Figures

Springer-Verlag
Berlin Heidelberg New York 1976

RAOUL A. ROBINSON, B. Sc., Dip. Ag. Sci., Dip. Trop. Agric.,
2, Balmoral Terrace, Trinity Hill, St. Helier, Jersey, C. I., United Kingdom

ISBN 3-540-07712-X Springer-Verlag Berlin · Heidelberg · New York
ISBN 0-387-07712-X Springer-Verlag New York · Heidelberg · Berlin

Library of Congress Cataloging in Publication Data. Robinson, Raoul A. 1928-Plant pathosystems. (Advanced series in agricultural sciences; 3) Bibliography: p. 1. Plant diseases. 2. Plant-breeding. 3. Plants—Disease and pest resistance. 4. Plants, Protection of. I. Title II. Series. SB 731.R 564 581.2 76-8914

Printed in Germany

Typesetting, printing and binding: Brühlsche Universitätsdruckerei, Gießen.

Foreword

One of the points clearly stressed in the beginning of this book is that the essential feature of any dynamic system is change and that, where there is change, there may also be growth and evolution. Plant breeding and plant protection have grown and evolved considerably during the past century; they have also witnessed several, important Hegelian changes. This book, by R.A.Robinson, is just such a change in scientific thinking. It is unique in presenting an entirely new insight to plant-parasite relationships, and in providing a practical guide for managing plant pathosystems for man's advantage in agriculture.

The author brings together for the first time in a holistic manner the various plant protection and breeding disciplines; he analyses their past limitations and deficiencies and throws useful new light on the nature of parasitism. From this he is able to gain a clear understanding of the functions of the various pathosystem components. On the basis of this understanding he then proposes practical ways for using these components to achieve and maintain the type of balance which is the basis of survival in any evolutionary system, including that of man himself. All this is done in a lively and elegant manner, using logic as the main driving force to elucidate and define entirely new concepts without obstruse mathematical or biochemical formulae.

The book is aimed mainly at the young scientist in plant protection and breeding who has had no part in the crop "boom-and-bust cycle" of the past half century, and who is looking for a new conceptual framework on which to orient his future career. It is also directed at the more mature, open-minded scientist who is willing to rest for a moment and consider his own work experience from a different scientific perspective.

Most important, this book is written for developing countries and, particularly, for those which are heavily dependent on subsistence farming and which, consequently, can afford neither the ill-effects of resistance breakdowns nor the sustained use of pesticide chemicals. To this end, the author has drawn heavily on his long experience in Africa to suggest a new plant-breeding approach fully within the scientific and technical capabilities of these countries. It is for this reason also that the Food and Agriculture Organisation of the United Nations has taken a world initiative in this field by launching a new International Programme on Horizontal Resistance. Contrary to T.H.Huxley's definition of the great tragedy of science—the slaying of a beautiful hypothesis by an ugly fact—we hope that this FAO Programme will serve to slay ugly facts with a beautiful hypothesis, which is contained in the following pages.

L. CHIARAPPA
Food and Agricultural
Organisation of the United Nations

Preface

To avoid possible misunderstanding, I must make four comments about this book:

1. It was written in various remote places which were all far from good library facilities. The bibliography is consequently inadequate and, given the choice between this and endless delays in publication, I chose what I considered to be the lesser of two evils. This means that I have made assertions of fact which are not supported by scientific references. This is not very important. It also means that many author 'credits' have been unjustly omitted. I apologise unreservedly for this and, should the book ever run to a second edition, I hope to correct this defect.

2. I must emphasise that the book is theoretical and speculative. This is deliberate as I regard speculation as a valid form of research. However, if it is to be good research, speculation must obviously be of a high quality, and this is a point which my readers must decide for themselves. Criticism concerning the quality of the theory and speculation will thus be welcome; but critics who dislike the book solely because it is theoretical must appreciate that it was never intended to be otherwise.

3. I have some hard things to say about certain aspects of the traditional disciplines of plant pathology and plant breeding. I consider the criticisms to be mild in view of how unholistic these disciplines have been in the past. No doubt, many readers will disagree with this opinion, as is their right. But I wish to emphasise that, in general, I have refrained from criticising individuals and that I hope that no one will feel personally affronted.

4. Having been a member of the field staff of the Food and Agriculture Organisation of the United Nations, I must state that nothing in this book is necessarily the official view of FAO and that FAO is in no way responsible for any imperfections which it may contain. On the other hand, if the book has merit, this is in no small measure due to my employment by FAO.

It is with great pleasure that I acknowledge valuable assistance from the following friends and colleagues, listed alphabetically:

L. Chiarappa, of FAO; R.B. Contant, of Nairobi University; T.J. Crowe, of FAO; N.A. van der Graaff, of FAO; W.C. James, of FAO; J. Kranz, of Giessen University; C.O. Person, of the University of British Columbia; C.J. Rossetto of Instituto Agronomico, Campinas, Brazil; H.I.A. Stoetzer, of FAO; J.M. Waller of the Commonwealth Mycological Institute; and R.K.S.Wood of Imperial College, London. I also wish to thank C.O.Person, C.J.Rossetto, N.A. van der Graaff, H.A.I.Stoetzer, J.Bruce, P.Dickinson and A.E.Bigornia for permission to use unpublished material. I need hardly add that I accept sole responsibility for any defects which may be present. My final, and perhaps, most important acknowledgment is that this book leans heavily on the writings of J.E. van der Plank who has done so much to clarify our ideas.

April, 1976 R.A.Robinson

Contents

Chapter 6 Horizontal Pathosystem Management

Chapter 7 Polyphyletic Pathosystems

Chapter 8 Crop Vulnerability

Chapter 9 Conclusions

Chapter 1 Systems

A pathosystem is comparable to an ecosystem. Indeed, a pathosystem is a subsystem of an ecosystem and is defined on the basis of parasitism. In particular, the term pathosystem emphasises that the whole subject of crop loss due to parasites is one system, and not several, entirely distinct systems as implied by the disciplines of plant breeding, plant pathology, entomology, nematology, and so on. These disciplines have tended to be separate, water tight compartments for far too long. Very broadly, breeders have studied the host and parasitologists have studied the parasites, and the areas of common ground have been remarkably limited. It will become apparent that the pathosystem concept, consisting of pathosystem analysis and pathosystem management, unites these disciplines into one cohesive new approach. The very existence of the word ecosystem has accelerated systems analysis in ecology (Watt, 1966; Patten, 1971, 1972); it is possible that the word pathosystem will do the same for the prevention of crop loss. However, before considering plant pathosystems, it will be useful to consider some of the wider aspects of the systems concept and its effects on science in general.

1.1 The Systems Concept

1.1.1 Origin of the Systems Concept

Although systems have been studied since the beginings of science, the systems concept is relatively new. Its origins probably lie in the field of communications research and information theory; its basis is the realisation that the most important fundamental in science is the pattern (Wiener, 1950). A pattern is an arrangement of units, and an essential feature of any pattern is that it is organised; the arrangement of the units is more important than the units themselves. For example, a word is a pattern, so is a message, a molecule, or a mathematical formula; a painting is a pattern in space, a tune a pattern in time, a game of chess a pattern in both space and time.

1.1.2 Definition of System

A system is a pattern of patterns, or more commonly, many patterns of patterns. In other words, we define a pattern as an arrangement of units but each unit is also a pattern. A pattern within a pattern is called a sub-system and a complex system has many levels of sub-systems. Equally, most systems are themselves units within a larger pattern which is then called a super-system. In biological systems, the pattern is often a population and the unit of the pattern an individual. Thus, an agricultural system is a population of individual crops, a crop a population of individual plants, a plant a population of individual cells and so on, down to the individual atomic particles.

Many systems are hierarchical, which means that there is a descending series of sub-systems, or ranks, each rank being qualitatively inferior but numerically superior to the rank above it. Military and taxonomic hierarchies are typical examples. Systems controls are often hierarchical also.

1.2 Properties of Systems

1.2.1 Structure and Behaviour

Considering a word as a typical pattern, it has the two basic properties of spelling and meaning and, characteristically, a small difference in spelling can lead to a very large difference in meaning. In most systems, spelling is called structure, and meaning is called behaviour. Systems analysis is the study of both systems structure and systems behaviour. Structure, being much easier to study than behaviour, is always studied first, but during the past century, the biological sciences have developed steadily from the study of structure to the study of behaviour. Modern science does not ignore structure, but studies it primarily for the better understanding of behaviour.

More loosely, we must recognise another relationship. The study of structure is essentially a study of facts, while the study of behaviour is more concerned with concepts and ideas. During the past century, there has also been a trend away from the extremes of Victorian empiricism towards a more balanced blend of both facts and ideas.

These relationships are broadly true of plant breeding and plant pathology. Gradually, we are moving away from an almost exclusive study of structure towards a study of behaviour while at the same time, it is to be hoped that we are becoming less biased in favour of facts and more interested in ideas. One of the many advantages of the systems concept is that it has clarified and, possibly, accelerated these trends.

1.2.2 The Holistic Approach

Closely parallel with the change of emphasis from structure to behaviour is the development of the holistic approach. This means that, figuratively and, occasionally literally, we are more interested in the wood than the trees, in the system than in the units of that system. Thus, in evolution, we now stress the effects of a multiplicity of selection pressures on innumerable micro-mutations in the gene pool, in place of the old concept of the survival of the fittest individual, such as the giraffe with the longest neck.

Traditional plant breeding for disease resistance has been unholistic, and has over-emphasised the host and under-emphasised the pathogen. It has concentrated on the individual host plant, the individual disease, the individual resistance mechanism, the individual chromosome and the individual gene. By being essentially Mendelian it has over-emphasised a few, prominent characters at the expense of the totality of survival values, the population and the gene pool.

Traditional plant pathology has also been unholistic. It has over-emphasised the pathogen and under-emphasised the host, and has studied the individual life cycle with a view to finding its most vulnerable point. Such studies lead to artificial disease control measures. In so far as resistance was studied, the single spore culture, the single detached leaf and the single resistance mechanism were stressed. Epidemiology was concerned with the mechanism of spore dispersal and the most simple ecological factors, usually considered in isolation, and only recently has epidemiology involved populations of both host and pathogen, their population dynamics and population interactions. The same has been broadly true of other disciplines studying crop parasites, such as entomology and nematology.

This also applies to the one small area in which traditional pathology and breeding really merged; the gene-for-gene relationship (3.2). The study of this relationship became a fashion, almost an obsession, until the whole picture of disease resistance in plants became distorted to the point of absurdity.

The results are obvious: we have stressed a few individual components of a highly complex system, but we have largely ignored the system itself. After some 70 years of "scientific" breeding of plants for disease resistance, we are no nearer to controlling many of those diseases than when we started. Ironically, we have been least successful with the diseases that have been the most intensively studied. We still suffer savage epidemics of wheat rust and many millions of acres of potatoes are still sprayed with expensive fungicides each year in order to control *Phytophthora infestans*, using a technique which is nearly 100 years old.

The function of systems analysis is to improve systems management. Our pathosystem analysis has been inadequate and defective, nor can we boast of our pathosystem management, which has often been pathosystem mis-management.

1.2.3 The Multi-disciplinary Approach

Another important development in science which has been greatly assisted by the systems concept, is the breakdown of the old university disciplines. The systems concept is essentially multi-disciplinary; a man who studies an ecosystem cannot be either a botanist or a zoologist, he must be both and more besides. The old perpendicular classification of scientific disciplines is consequently disappearing, but because no man can study everything, it is being replaced by a lateral classification. (The more commonly used terms vertical and horizontal are avoided in this context for obvious reasons). This means that all the relevant, perpendicular disciplines are studied but mainly at one system level.

With the development of the systems concept it is also becoming apparent that the old university disciplines have common ground and a common terminology. The principles of systems analysis and systems management are the same in all the old disciplines and, equally, incorporate all the old disciplines. It seems that, quite suddenly, and even unexpectedly, the old disciplines are being unified and that we can begin to think in terms of "science" as a whole.

Equally significant is the disappearance of the old and invidious distinction between pure science and applied science in which the latter, by implication, was impure. As Medewar (1967) has pointed out, this distinction leads to the dire equation: "useless = good; useful = bad". With the gradual unifying effects of the systems concept, the boundary between pure and impure science is becoming so blurred that the distinction is now largely meaningless.

Traditional plant breeding and pathology are still perpendicular disciplines; a proponent of either is apt to be separated from the other by an intellectual boundary which leads to isolation and, quite often, downright hostility. If he attempts to cross that boundary, he is likely to be met with an attitude which tells him, no doubt politely but very firmly, to mind his own business, and to stick to his own subject. This is the intellectual equivalent of the territorial imperative in animals. It is remarkably primitive. However, we recognise that the study of human behaviour, as opposed to human structure, is a very recent development. We also take comfort from the fact that scientists are quicker than most to recognise the more primitive aspects of their behaviour and to adjust accordingly.

1.2.4 Dynamic Systems

A system may be static or dynamic. A taxonomic system is static; it can be revised but it is then a different system. There is, of course, a strong conceptual and linguistic relationship between the words system and systematics.

The solar system is a dynamic system and, like any dynamic system which endures, is a stable system. This stability is more commonly known as dynamic equilibrium and, in biological systems, is often called systems balance. If systems balance is lost, the system as a whole becomes unstable and even self-destructive.

The essential feature of a dynamic system is change, which can occur at any systems level in a multi-level system. A dynamic system also has variable and non-variable elements; the speed of rotation of a cog-wheel is a variable element, but the number of cogs on the wheel is a non-variable element. Variables are subject to the mathematics of extremes; they have a minimum, an optimum and a maximum. In the evolutionary system, variables are usually known as survival values; many different survival values occur at all systems levels and we thus refer to evolution as a multi-variate system.

Change implies time and rates of change both differ and are themselves variable. We consequently recognise the multi-rate system, which may involve time scales ranging from those of atomic particles to geological time. Equally, because change implies time, every dynamic system has a history which must be taken into account in systems analysis. The units of history in the evolutionary system are fossil strata in which a few individuals of a population were preserved with their basic structure unchanged. But other time scales are also involved; the units of history in a plant disease epidemic, for example, are usually measured in days. Systems history is usually analysed mathematically with recurrence formulae expressing the state of the system at time $(t+1)$ as a function of the state of the system at time t. This is the basis of the mathematical analysis of epidemics.

1.2.5 Open Systems

In thermodynamics, a closed system is one which cannot exchange energy across the boundaries with its external environment. The second law of thermodynamics states that, in a closed system, entropy must increase; that is, all temperature gradients must decrease until there is a uniform temperature throughout the system.

By analogy, the concept of entropy can be applied to patterns and systems, although it is convenient to refer to its converse, which is called negative entropy. The evolutionary system and, hence, all biological systems, are open systems, gaining radiant energy from the sun and converting it to the chemical energy which is the basis of life, resulting in the increase of gradients of negative entropy. If the supply of solar energy were cut off, all life would cease quite rapidly in the perpetual darkness and cold. Typically, the most highly evolved life, with the steepest gradients of negative entropy, would cease soonest, and the least evolved would endure, no doubt quite dormant, longest.

In systems terminology, negative entropy can be defined as a mathematical improbability of pattern and, thus, an improbability of structure and behaviour. Consider a gradient of negative entropy in a simple system based on letters and words: a high improbability of pattern represents high negative entropy, a high randomness of pattern represents low negative entropy. Improbability of pattern is equivalent to organisation. The least organisation is an even, paint-like spread of printer's ink over the page. Greater organisation occurs with a random spatter of ink; greater still with a random printing of letters; greater still with a meaningful arrangement of letters, words and sentences. H. J. Robinson (1975) has commented that the concept of negative entropy possibly provides the only objective, scientific basis for human values and that all high human values have high negative entropy.

Improbability of pattern can be illustrated another way. Given an alphabet of twenty six letters, a finite number of meaningful words is possible. In the multi-decision process of writing, the choice of a particular word at any one time interval has a fairly high mathematical probability. At the next systems level, given the finite number of words, a finite but vastly greater number of meaningful sentences becomes possible, the choice of any one sentence being thus more improbable than the choice of any one word. The sentence has a greater negative entropy than the word, but at the next systems level, the book, the total number of possible patterns is beyond computation. Any book is highly improbable compared with the vast number of similar books which might have been written in its place but were not.

The same is true of the evolutionary system. If we take the atom as the unit of pattern, the numbers of multi-molecular cells and multi-cellular animals which could have evolved is mathematically vast. A human individual thus presents a high improbability of atomic pattern when compared with, say, a crystal of comparable size.

From negative entropy, we derive the systems concept of growth, which occurs in many dynamic, open systems at all systems levels. The rate of growth is related to the system level. Evolutionary growth is slow; the growth of a new species is faster; the growth of an individual within that species is many times faster; the growth of a cell within that individual is faster still. Growth may be exponential; that is, its rate of increase is itself increasing, and exponential growth leads to a population explosion.

Implicit in the concept of growth is the concept of reproduction, which means that an entire pattern is re-formed, often repeatedly and at any systems level. Printing is a typical example; so is the growth of a chain of retail shops. Reproduction is an essential feature of the evolutionary system and it occurs at all systems levels, from the DNA strand to the population of a species.

Growth can also be negative; a loss of negative entropy. Negative growth can also be irreversible, leading to the extinction of the sub-system. Extinction is essential in the evolutionary system, at all systems levels; it is the converse of reproduction and occurs with the same frequency. At the sub-system levels of the individual organism and cell, the process of extinction is usually two-fold. First there is a loss of behaviour, which is death; then there is a loss of structure, which is decomposition. Decomposition, in its turn, makes new growth and new reproduction possible, hence the concept of food chains.

An important property of both positive and negative growth is the Hegelian change in which a small difference in degree produces a difference in kind. The critical state at which a small quantitative change leads to a qualitative change is often called a point. Thus, the melting point, boiling point, thermal death point, dilution-end point, break-even point, flash point, bursting point, and so on. In biological systems, this point is often called a threshold value.

The growth of the evolutionary system is a long history of Hegelian changes. Man represents the latest Hegelian change and, as we shall see, is now changing the evolutionary system itself with some entirely new factors.

Another feature of growth is competition and its converse, cooperation. Competition is a key feature of the evolutionary system but it has been much overemphasised. When the Darwinian theory was new, for example, the apparent necessity for competition was used to justify man's exploitation of man. It was not appreciated that competition must be controlled and that unbridled competition leads to a loss of systems balance which can be highly destructive. Systems balance and the avoidance of a self-destructive loss of balance can only be maintained when competition and cooperation are approximately equal.

Parasitism is a special case of competition; one species, the parasite, survives at the expense of another, the host. But systems balance is crucial; if the parasite is too competitive, the evolutionary survival of the host will be impaired, and if the host becomes extinct, the parasite will also become extinct. It is for this reason that there is an absolute limit to the parasitic ability of parasites. Without such a limit, "cooperation", in the systems sense of the word, systems balance would be lost and the system would become self-destructive. It follows that in evolution, all surviving, natural parasitic systems can only be balanced, otherwise they would not have survived.

1.2.6 Systems Control

A dynamic system can remain stable only if it retains systems balance. The maintenance of systems balance is achieved by systems control. Control involves communication between the units of a pattern; its study is called cybernetics. The control may originate from outside the system which is then a dependent system, or it may originate within the system which is then an independent or autonomous system. In an autonomous system, communication between the units of a system is reciprocal, as with a thermostat which reacts to changes of temperature within the system. This reciprocity is called feedback.

The control itself can be either autonomous or purposeful and deterministic. This, as we shall see, is the definitive characteristic of man-made systems. Plant pathosystems can be divided into two quite distinct categories on the basis of their systems control. A natural, or wild plant pathosystem is autonomous, the autonomous control being primarily due to communication between the three basic components of the pathosystem; the host, the pathogen, and the environment, for which reason the natural pathosystem is sometimes called the disease triangle. An artificial or crop pathosystem has a fourth component which is man, and is consequently called the disease square. Man has had a profound influence on the other three components; cultivars differ from wild hosts, cultivation differs from wild ecosystems and the pathogen population can be artificially controlled. The disease square thus differs from the disease triangle in its element of purposeful, deterministic control.

Each step in systems control is called a decision, and a dynamic system clearly involves many multi-decision processes. There are three categories of multi-decision process. A Markov chain is a multi-decision process in which every decision is entirely random, such as is produced by spinning a coin or throwing dice. A game of roulette is a Markov chain and gamblers who think otherwise, as those who only bet on black when red has appeared four times in succession, and then double their bets to restore their losses, deserve to lose. A stochastic decision process is one in which the decisions are still random but in which each decision is related, on a probability basis, to the state of the system at the time of the decision. For example, at a particular stage of the game, a chess player might have one hundred possible moves to choose from. If he put all these moves in a hat and drew one out at random, this would be a random decision which, however, was related on a probability basis to the state of the game at that time. A purposeful or deterministic decision process occurs in the normal game of chess, in that each decision in such a multi-decision process is non-random; it is made in accordance with the best interests of the pre-determined objective of winning. In a complex system, there may be various mixtures of Markov chains, stochastic decision processes and deterministic decision processes.

Returning to the two categories of plant pathosystem, it is clear that a natural pathosystem (disease triangle) is normally a stochastic decision process. Whether or not a particular infection site becomes infected is a random event, but it is also related, on a probability basis, to the state of the system at that time. In the crop pathosystem (disease square) there is the additional factor of deterministic control which, for example, might or might not decide on the application of a fungicide.

Most controls show fluctuations in their effectiveness; one kind of fluctuation is called hunting and is an acceptable variation each side of the optimum. The term is used mainly in mechanical systems and hunting occurs typically in automatic speed control and navigational systems. However, the phenomenon of hunting is a common feature of any autonomous system, occurring for example, in the autonomous control of mammalian body temperatures.

A special case of hunting occurs in parasitic systems. If the population of the parasite becomes too large, the population of the host decreases. The parasite population then also decreases and as a result the host population increases. Such hunting around the optimum population sizes is a feature of all natural parasitic associations.

Another fluctuation in control is cyclical. Many biological cycles are diurnal and are the result of the earth's rotation around its own axis, others are seasonal and are the result of the earth's rotation around the sun. The main seasonal effects in the temperate regions are those of temperature and day-length; in the tropics, the seasonal effects are primarily those of alternating wet and dry seasons. The rotation of the moon around the earth has profound significance in tidal ecosystems.

Other biological cycles, however, are largely independent of solar system cycles. In particular, the reproductive cycles of micro-organisms are usually too rapid to be affected, except in the most general way.

An epidemic cycle is the positive growth of a parasite population from minimum to maximum, followed by a negative growth to minimum again, and is normally seasonal, except in the wet tropics which are permanently warm and humid. A disease cycle, however, from host infection to pathogen reproduction to host infection, is not necessarily seasonal. In a polycyclic disease, such as wheat rust or potato blight, there are many disease cycles to each epidemic cycle, whereas in an oligocylic disease, such as a wilt, there are only a few disease cycles to each epidemic cycle and, in a monocyclic disease, such as the cereal smuts, the disease cycle and the epidemic cycle coincide.

The host cycle can also profoundly affect an epidemic, mainly by the removal of the host tissue, occurring with the death of annual plants and the leaves of a deciduous tree and, in agriculture, with the harvesting of the crop, when the epidemic cycle comes to an abrupt halt. Other epidemic cycles decline more gradually with the change of season.

These various aspects of control, whether autonomous or deterministic, involving feedback and innumerable interactions in a complex, multi-decision process swayed by cyclical and other fluctuations, are all factors in the stability of a dynamic system. A stable system has the property of resilience; it is well buffered and can exhibit fairly wide swings from the optimum and yet recover. This property of maintaining a stable optimum is called homeostasis, a term first used by Cannon (1939) with respect to the self-regulating devices in mammalian physiology. It has subsequently been used in many systems and we now speak of genetical (Lerner, 1958), political, military, ecological and other forms of homeostasis. An example of homeostasis which is of particular relevance in plant pathosystems is the Person model (5.4) which is designed to demonstrate the stability of the horizontal pathosystem when the populations of both host and parasite are genetically flexible.

1.2.7 Man-made Systems

We have seen that the evolutionary system was autonomous and that it involved many Hegelian changes. The most recent of these changes occurred a few million years ago. Small quantitative changes in intelligence, manual dexterity and vocal ability produced in man an entirely new evolutionary phenomenon. The main component of this phenomenon involved an increase in the effectiveness of communication and this led to the development of a species memory, as opposed to the individual memory. Memory became possible at a new and higher systems level. This species memory can grow indefinitely because the species endures for geological time while the individual memory endures for only one life-span, and its growth both assisted and was assisted by the parallel development of tool-making (as opposed to tool-using). Both developments showed exponential growth, often accelerated by Hegelian changes which occurred with increasing frequency and magnitude. This growth is called cultural development and one of its features was the appearance of artificial, man-made systems which differ from natural systems in their deterministic control. Examples include legal systems, political systems, military systems, economic systems, educational systems, industrial systems, urban systems, mechanical systems, electrical systems, and so on.

Within the overall evolutionary system, the most important result of this development was the appearance of man's entirely new, deterministic control of evolution itself. Primitive man was a combined food gatherer and hunter, depending entirely on wild populations of both plants and animals, and he himself was a component of a natural ecosystem. One of the first steps in man's cultural development was the discovery of cultivation; seeds of useful plants were sown and the resulting plants were protected, the most immediate effect being an increase in the food supply. This meant both that a given area of land would support a larger human population and that a proportion of that population was spared from food procurement and became available for other aspects of cultural development, such as education, tool-making and so on. Man thus ceased to be a mere component in a natural ecosystem; he was creating his own, artificial system in which the distinguishing characteristic was the new element of purposeful, deterministic control. It is relevant that the common feature of all religious systems is a postulation of a deterministic control of the natural events in an autonomous system, and it is only quite recently that the complete autonomy of natural systems has been recognised. This is relevant to the now out-moded concept of teleology, the doctrine of design or purpose in a natural system. A teleological argument implies a purposeful, deterministic control; for example, berries are red in order that birds may know they are ripe, and to avoid teleology we must argue that redness is a natural survival value which increases the probability of seed dispersal. Teleology was anathema to the early proponents of Darwinism but there is no need for us to re-fight those old battles.

Possibly the most significant feature of plant cultivation was that seed was saved from each crop to plant the next crop, and that this seed could be selected. An entirely new, deterministic control of evolution itself appeared; a control process which is called domestication.

Domestication means that certain selection pressures are reduced and others increased. The change of emphasis is imposed by man, hence the term artificial selection. An essential feature of cultivation is that the crop is protected from much natural competition. There is then negative selection pressure for survival values which contribute to competitiveness in a natural ecosystem. At the same time, man exerted positive selection pressures for other survival values which contribute to both the quantity and the quality of the harvestable product.

One of the effects of this deterministic control of evolution is a reduction in time scales. The autonomous, natural evolution of a new species requires a few million years, while the deterministic domestication of a new species, such as wheat or maize, requires a few thousand. However, we must make a distinction between deterministic control, which is largely undeliberate and ill-informed, and control which is both deliberate and well-informed. With modern techniques and knowledge, we can re-synthesize wheat or maize in a few tens of years, thus making a further thousand-fold reduction in the time scale. High levels of deterministic control thus permit an artificial evolution which is about one million times faster than natural evolution. This is the real evolutionary significance of the Hegelian change called man.

Eventually, man will dominate natural evolution entirely. We are already at the stage in which the man-eating carnivores would be extinct but for the fact that we choose to preserve them, and the same will soon be true of smallpox and other human parasites. Eventually, this will be true also of all natural ecosystems which will survive only as museums because man so decides.

Both tool-making and tool-using have increased man's natural abilities by many orders of magnitude. This is seen in machines which replace muscles, in terms of both strength and precision. Machines have also improved transport, both of goods and man himself, far beyond the natural limits of carrying and walking. Communication has been vastly increased by the sending of messages, the recording of which has been improved, first by writing, then by printing and now by computer storage and retrieval. No individual memory can retain more than a fraction of the current species memory which continues to grow exponentially. Finally, by the use of computers, the thinking function of man's brain is also being increased, not only in terms of previously impossible mathematical calculations but also, it seems, in terms of generating new ideas which man's brain might never have produced unaided.

Computers and communication between computers are the latest man-made systems. They also represent the most recent, and possibly, the greatest Hegelian change in cultural development. To quote only one relatively mundane example which, however, is relevant to pathosystem analysis, they permit previously impossible multiple regression analyses. Van der Plank (1975) comments: "Future historians may well come to regard Schrodter and Ullrich's (1965) introduction of multiple regression analysis into the epidemiology of plant disease as one of the milestones in plant pathology". It may well, indeed, result in new ideas which our brains could not have produced unaided. However, computerised analysis and simulation of epidemics, which have been reviewed by Kranz (1974a) and van der Plank (1975), are beyond the scope of the present book

1.2.8 Social Systems

We have seen that cultivation and domestication lead to increases in agricultural productivity, a prerequisite for the growth of other social systems which require a proportion of the population freed from the necessity to procure food. In a developed country, some 5% of the population are engaged in agriculture, whereas in an undeveloped country, as few as 5% may be engaged in social systems other than agriculture. This is due primarily to low productivity but also demonstrates that social systems are inter-related. There are three factors in low agricultural productivity: the farmers are ignorant, not because they are un-intelligent but because they are uneducated, having been denied access to the species memory. But the educational system cannot develop until higher productivity releases individuals from agriculture. Low productivity and lack of education also restrict the number of individuals available for tool-making and industry, so that the products of industry, such as good roads and transport, tractors and fertilizers are largely lacking. The farmers are also cultivating plants which are relatively primitive and unproductive, often described as subsistence cultivars because they enable the farmer and his family to subsist but little more.

The mention of subsistence cultivars brings us to the end of this rapid survey of the systems concept. In the spectrum of artificial selection, subsistence cultivars are approximately half-way between unselected, wild plants and highly selected, highly productive, modern cultivars. The evolutionary survival of wild plants is not impaired by their parasites. Subsistence cultivars may have low yields but they can be grown without serious loss from pests and diseases, in spite of artificial control methods which, at best, are rudimentary. Yet the most productive of modern cultivars often cannot be grown at all without an extravagant use of that group of chemicals called pesticides.

We must now enquire, therefore, whether we can combine the high pest and pathogen resistance of subsistence cultivars, or even wild plants, with the high yields and quality of modern cultivars. If we are to do so, it seems that our approach must be holistic, multi-disciplinary and behaviourial, and must involve systems analysis with a view to improved systems management. Our pathosystems analysis must, above all, compare the artificial pathosystem (the disease square) with the natural pathosystem (the disease triangle), and our pathosystem management must aim at the restoration of systems balance, which is so prominent in the natural pathosystem and so conspicuously absent in the artificial pathosystem.

1.2.9 Plant Pathosystems

As with the term ecosystem, the term pathosystem can be given whatever geographical boundaries are most convenient; its conceptual boundaries can also be defined as convenient. In its widest sense, a plant pathosystem is a sub-system of the ecosystem and is defined by the phenomenon of parasitism, which involves all the hosts of the ecosystem, and all the parasites. A crop pathosystem involves only one crop species, whatever its geographical boundaries, but it involves all the parasitic agents of both pre-harvest and post-harvest loss within that crop. However, both disorders (which can be defined as non-parasitic malfunctions due to an excess or deficiency in environmental factors such as wind, heat, light, water or nutrients) and competition from weeds, are outside the conceptual boundaries of the pathosystem and belong to the wider concept of the ecosystem.

This book is designed to be no more than an introduction to the pathosystem concept, and to pathosystem analysis and management. Relatively few crops are mentioned and these have been chosen by the criterion of their useful illustration of basic principles. The same is true of the parasites where the emphasis is intentionally on a few carefully chosen plant pathogens, although other parasites are mentioned when relevant.

1.3 Systems Analysis

We analyse systems because we want to understand them, as systems. We need to understand them because we want to manage them, as systems; above all, we want to manage them effectively.

1.3.1 Scope

Systems analysis can range from the intensely theoretical to the essentially practical, the latter being most familiar in the popular mind in systems such as commercial airlines, military supplies, space travel, computer programming, telephones, town planning and so on. This book is primarily concerned with theoretical systems analysis, although excursions are made into the practical for purposes of illustration.

Systems analysis can also range from the most general to the most detailed. General analysis concerns fundamentals which apply to many systems; detailed analysis concerns the minutiae of one particular system. This book is primarily about the general, although some detail is included for illustration.

1.3.2 Plant Pathosystem Analysis

About half of this book is concerned with pathosystems analysis. It will become apparent that pathosystems analysis is holistic, that it studies pathosystem structure only for a better understanding of pathosystem behaviour. It will be found that certain behaviourial characteristics are common to many pathosystems. For this reason, the pathosystem analysis is more concerned with basic principles, concepts and ideas than with the actual characteristics of individual plant pests and diseases. Anyone who requires a catologue of facts concerning these parasites should consult reference books, of which there are many. Indeed, a basic knowledge of the plant sciences is assumed throughout, pathosystem analysis being also multi-disciplinary. Above all, it is an attempt to apply the systems concept in a field in which it has been neglected for far too long.

1.4 Systems Management

We undertake systems analysis in order to improve our systems management, which involves the deterministic control of a system in order to achieve predetermined objectives in the most effective way possible.

1.4.1 Plant Pathosystem Management

Management of the crop pathosystem is concerned with reducing the crop loss due to parasites, but as the term systems management implies, crop-loss prevention must be at its most effective. This means that it must be as cheap, as complete, and as permanent as possible, it must also be achieved without any loss of crop yield or quality, and it must not interfere with the wider aims of crop husbandry, such as monoculture in both time and space, the uniformity of the crop and the crop product.

Cheapness means that the use of pesticides must be reduced to a minimum. On a world scale, millions of tons of an irreplaceable asset such as copper have been irretrievably dispersed over vast areas of crops. Other pesticides are more economical of such basic assets but they absorb manufacturing potential which, eventually, can be more effectively utilised. At the farmer level, pesticides cost money, effort and technical expertise. It must not be inferred that pesticides are necessarily bad; compared with the crop losses they prevent, they are both economic and essential, but they represent poor pathosystem management if the same or even better crop-loss prevention can be achieved without them. In other words, the development of pesticide use was a major improvement in pathosystem management; but we can do better yet.

Complete pathosystem management means two things, that the crop loss due to any one parasite species must be reduced to a negligible amount, and that this is true of all the locally important parasites; hence the concept of comprehensive resistance (9.2).

Permanence in pathosystem management means three things: pesticides usually provide a control of parasites which is remarkably temporary in that they must be frequently reapplied, another reason why they represent pathosystem mismanagement; most breeding for resistance has involved temporary resistance, and we must work with resistance which is permanent, at least during the foreseeable agricultural future. Some parasites will never be locally important because they lack epidemiological competence (6.4). This is not true of others which are the cause of crop vulnerability, discussed in Chapter 8, and which must be taken into account if the effectiveness of our pathosystem management is to be permanent.

There is a widely held but mistaken belief that high resistance to parasites is correlated with low crop yields and quality, which is refuted (9.1). In the past, plant breeding for parasite resistance has been incompetent; it has been unholistic and has generally emphasised resistance to a single parasite due to a single and prominent resistance mechanism, preferably inherited by a single major gene. It has also emphasised Mendelism, pedigree breeding methods and back-cross programmes which are both wasteful of research resources and are without doubt the techniques least likely to produce resistance which is both comprehensive and permanent.

There is another popular but mistaken belief that crop uniformity leads to a peril called the "genetic vulnerability" of crops, which can also be refuted (8.2). Crop uniformity is necessary if we are to obtain the monoculture and bulk handling of crop products necessary to a highly productive agriculture. It is also safe, provided that the crop pathosystem is not grossly mismanaged.

At this point the diehards of the traditional disciplines will probably refuse to read further. Let it suffice that all the claims just made for pathosystem management will be substantiated later in the book. One qualification must, however, be clearly stated, that some crop species are easier to manage than others. There is, in fact, a spectrum of crops based on the ease of their pathosystem management. The most difficult are very difficult indeed, but even this does not represent impossibility.

Perhaps unexpectedly, the easiest crop to manage is sugarcane, and its pathosystem management is at its best in Hawaii. Apart from the chemical protection of seed setts, once every fourteen years, the Hawaiian cane crop is grown without the use of pesticides and has no serious crop loss due to parasites. This prevention of crop loss is both comprehensive and permanent. Hawaiian cane yields are the best in the world and are as high as 8 tons of sucrose per acre per annum even though sugarcane is a crop of great genetic uniformity in both time and space. Given good pathosystem analysis and competent pathosystem management, we should be able to do the same with many other crops and in many other areas.

At this point it is perhaps relevant to inquire not how we should manage the pathosystem, but who should manage it. It is perhaps relevant also to define the pathosystem manager, who is neither plant breeder, plant pathologist, nor entomologist, but all of these things and more besides, a new kind of plant scientist who is primarily interested in the pathosystem as a system, and whose aim is the restoration of pathosystems balance.

If it is successful, this new approach will show a characteristic growth curve which will be closely comparable to that of an epidemic, with positive and exponential growth to a maximum, followed by negative growth to a minimum. The increasing positive growth will be due to success; successful pathosystem management leads to the permanent solution of pathosystem problems. It is consequently self-eliminating and the negative growth will be a measure of its overall success. Eventually, there should be no work left for it to do.

However, any individual who is concerned about the loss of a career need have few fears. The successful completion of our overall pathosystem management may be achieved by the time that present-day undergraduates have reached retiring age; it is unlikely to occur any sooner.

Chapter 2 Plant Pathosystems

2.1 Definitions

2.1.1 The Terms "Vertical" and "Horizontal"

Van der Plank (1963, 1968) postulated that all disease resistance in plants can be classified into one of two categories which he called "vertical" and "horizontal". Vertical defines a type of interaction between host and pathogen, horizontal defines another type of interaction. No other kind of interaction is possible, in spite of various complications due to mixtures of both interactions, and interactions with other factors such as those of the environment.

It will become apparent that the value, significance and precision of these two terms is now far greater than their author perhaps intended or even realised, which is a measure of his brilliance. Originally, the terms were used only to qualify two epidemiological categories of disease resistance in plants; now they can also be used to qualify resistance to pests and other parasites. They can also be applied to other protection mechanisms such as fungicides and insecticides, to the mechanisms of both resistance and protective chemicals, to the two categories and many mechanisms of pathogenicity and parasitism; to the genes and genomes governing the inheritance of both resistance and parasitism, to the sub-systems of a pathosystem and to populations of both the host and its parasites.

It is clear that vertical and horizontal can be used in many different contexts, including agricultural, genetical, plant pathological, entomological, biochemical, epidemiological, histological, ecological and other disciplines. The terms are thus multi-disciplinary; they are, above all, systems terms, describing a concept, and for this reason they must be conceptual or abstract terms. All other terms which are approximately synonymous with vertical and horizontal are literal terms with a descriptive function which may be very precise in one context but cannot be precise in all contexts. A genetical term, for example, is meaningless and positively misleading in a histological or an epidemiological context, however precise it may be genetically. A further disadvantage of descriptive terms is that they can be ambiguous even in one context. For example, the word "specific" has several different connotations due to its close relationship with species, special, specified and its use in physics to qualify heat and gravity. When applied to resistance, specific is taken to mean a differential interaction (2.4) but, as we shall see, not all "physiologic specialisation" is due to vertical resistance, nor is the term appropriate when applied to incomplete vertical resistance (3.1), which is quantitative resistance.

Unfortunately, the terms vertical and horizontal have proved unpopular, probably for the very reason that they are non-descriptive. The use of inadequate, descriptive synomyns has been so inappropriate and so widespread in recent scientific literature that both our terminology and many of our ideas are in a state of confusion. At their worst, some scientific comments (which, in charity, need not be quoted) make one despair of those two multi-decision processes called thinking and writing. Accordingly, the first part of this Chapter is concerned with definitions.

Ackoff (1971) has commented that definitions are like surgical instruments; they become dull with use and require frequent sharpening and eventually, must be replaced. The following definitions have been sharpened for the purposes of this book; if the book is to be understood, the definitions must also be understood.

2.1.2 Origin of the Terms

The terms vertical and horizontal are derived from van der Plank's (1963) two classic diagrams which are shown, slightly modified, in Figures 1 and 2. These diagrams show the amounts of blight which various potato cultivars suffer when exposed to sixteen different races of *P. infestans.* In Figure 1, the top diagram shows a potato cultivar which possesses the single-resistance gene R_1. It is susceptible to all blight races which possess the pathogenicity gene v_1 and completely resistant to all races which lack this gene. The bottom diagram shows a potato

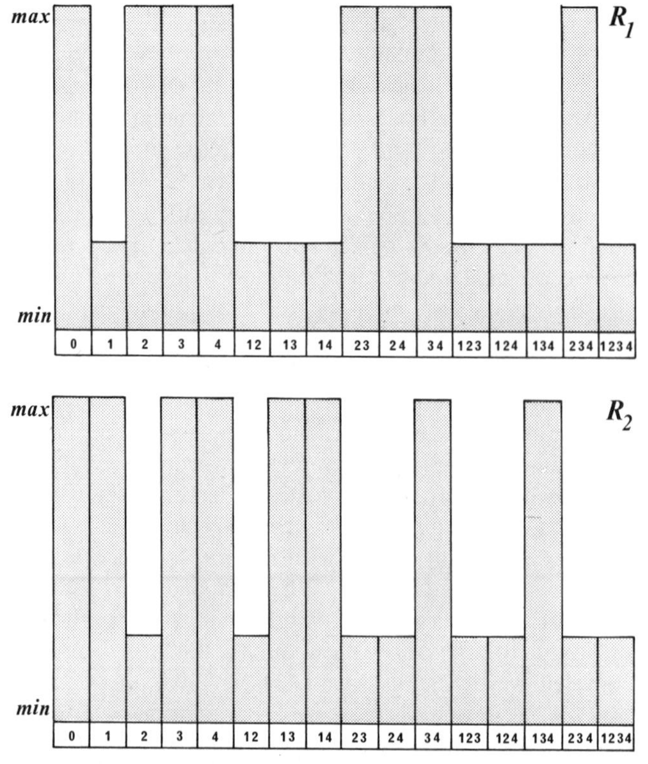

Fig. 1. Origin of the term "vertical"

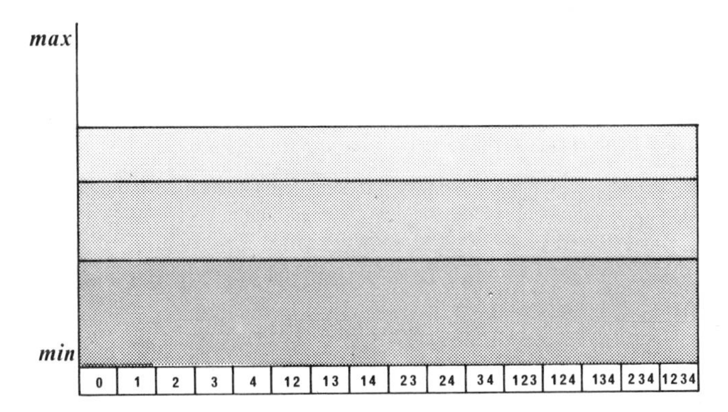

Fig. 2. Origin of the term "horizontal"

cultivar which possesses the resistance gene R_2 and which is thus susceptible to all races with the pathogenicity gene v_2 but resistant to all others. When the two diagrams are compared, it will be seen that differences between the two potato cultivars are parallel to the vertical axis of the diagram and this kind of resistance is accordingly called vertical resistance. Vertical resistance operates against some races of the pathogen and not others.

Figure 2 shows several potato cultivars which possess no R-genes at all. They are consequently susceptible to all the blight races but some are less susceptible or, conversely, more resistant than others. It will be seen that differences between these cultivars are parallel to the horizontal axis of the diagram and this kind of resistance is accordingly called horizontal resistance. Horizontal resistance operates equally against all races of the pathogen.

In the past, vertical resistance has been called differential resistance, field resistance, field immunity, hypersensitive resistance, major gene resistance, qualitative resistance, R-gene resistance, race-specific resistance, racial resistance, specific resistance. Horizontal resistance has been called field resistance, general or generalised resistance, minor gene resistance, multigenic resistance, non-hypersensitive resistance, non-racial resistance, non-specific resistance, partial resistance, polygenic resistance, quantitative resistance, quantitatively inherited resistance, race-non-specific resistance, relative resistance, residual resistance, tolerance and uniform resistance. All these terms have literal connotations and a descriptive function. They can be very precise in the appropriate context and they may be retained for use in that context, but they are inappropriate when used out of context and, for this reason, they cannot be regarded as synonymous with vertical and horizontal respectively.

2.2 The Agricultural Context

2.2.1 *Vertical and Horizontal Defined*

From a farmer's point of view, the essential feature of vertical resistance is that it is usually (but not necessarily) temporary resistance. Its effects are likely to be lost when a new race of the pathogen appears, when it is then said to breakdown.

Strictly, it is the effectiveness of the resistance which breaks down, the resistance itself is unaltered—it is the pathogen which has changed. In conversation, however, we speak of the breakdown of the vertical resistance in the same way as we say "the sun has come out" when we mean that the clouds have moved away, and such conversational usage is entirely legitimate provided that no ambiguity results.

Vertical resistance normally confers complete protection while its effectiveness lasts. In spite of exceptions, a useful working rule is that, in agriculture, vertical resistance provides a complete but impermanent control of a disease.

The completeness of vertical resistance makes it agriculturally popular. A cultivar with a "new" vertical resistance is thus increasingly cultivated in a "boom" of popularity. But the impermanence of vertical resistance leads to its breakdown and the "busting" of that cultivar. The repetition of this phenomenon has been called the "boom-and-bust cycle of cultivar production", and is the most important agricultural feature of vertical resistance. It means that the breeders are continuously employed producing a series of new cultivars in an attempt to "keep one jump ahead of the pathogen". With such repetitive breeding, it is difficult to produce cultivars with improvements in other qualities.

Because horizontal resistance operates equally against all races of the pathogen (Figs. 2 and 4), it does not break down due to changes in the pathogen population, but is permanent resistance, at least in the foreseeable agricultural future. However, it is usually inherited polygenically and is then quantitative in its inheritance, mechanisms and epidemiological effects. Compared with vertical resistance, therefore, a useful agricultural rule is that horizontal resistance provides an incomplete but permanent control of a disease. As with vertical resistance, however, there are some notable exceptions to this rule.

The permanence of horizontal resistance means there is no boom-and-bust cycle of cultivar production. Breeding for horizontal resistance is thus cumulative, a good cultivar being replaced only with a better cultivar.

2.2.2 Protective Chemicals

From a farmer's point of view also, any pesticide which protects his crops from parasites may be described as vertical or horizontal. The effectiveness of many insecticides, fungicides, rodenticides, etc, is temporary, because the parasite produces a new race unaffected by the chemical in question; these are vertical pesticides. Conversely, the effectiveness of other pesticides is permanent, at least in the foreseeable, agricultural future. The best known example is Bordeaux Mixture, a horizontal pesticide which has been in use long enough for us to be confident of its permanence.

2.3 The Genetical Context

2.3.1 The Manner of Inheritance

Both host resistance and parasitic ability are inherited characters and the manner of their inheritance is of concern to geneticists. Here we must anticipate and comment that the vertical pathosystem probably always involves a gene-for-gene relationship (3.2). Consequently it is probably safe to assume that vertical resistance and vertical parasitism are always inherited oligogenically. Conversely, polygenically inherited resistance and parasitism can only be horizontal, but because oligogenically inherited horizontal resistance is known, the converse relationships are not true; not all horizontal resistance is polygenic and not all oligogenic resistance is vertical.

We must now discuss three sources of confusion.

2.3.2 *The Fallacy of the Undistributed Middle*

This is the fallacy of "all but not only" and it leads to the classic false syllogism: "all tables have four legs; this dog has four legs; therefore, this dog is a table". Maybe all tables do have four legs, but not only tables have this property.

This particular fallacy is common almost beyond belief in the scientific literature on disease resistance in plants. We have just seen that all, but not only, vertical resistance is inherited oligogenically, all but not only polygenically inherited resistance is horizontal. As this book proceeds, many other similar fallacies will be noted. The reason for the frequency of the fallacy is distressingly simple. The authors concerned have insisted on using literal, descriptive terms out of context.

2.3.3 *Qualitative and Quantitative Characters*

Oligogenically inherited characters are normally expressed qualitatively; they are either present or absent, or they are present to the extent of a clear-cut, vulgar fraction. Polygenically inherited characters are normally expressed quantitatively; they can vary in all degrees of difference between a minimum and a maximum. This has led some authors to use the term "qualitative resistance" in place of vertical resistance, and "quantitative resistance" in place of horizontal resistance. Once again, these are literal terms with descriptive connotations. Not only vertical resistance is qualitative and not only horizontal resistance is quantitative.

2.3.4 *The Relevance of the Manner of Inheritance*

The number of genes governing the inheritance of resistance and parasitism is of relevance primarily to geneticists to decide questions such as the most appropriate plant breeding technique, but apart from the incomplete relationship already noted, the number of genes is irrelevant. This has not prevented various authors from assuming that there is a continuous spectrum from few genes to many genes and that, consequently, there is a continuous spectrum from vertical resistance to horizontal resistance.

Vertical and horizontal resistance are entirely different characters whose inheritance is controlled by entirely different genes. It is not a question of how many genes control the inheritance, but of which genes do so.

2.3.5 *Protective Chemicals*

When the terms vertical and horizontal are used to describe protective chemicals, the genetical relevance is clearly confined to the parasite only.

2.4 The Epidemiological Context

2.4.1 Populations

The first essential of epidemiology is that it concerns populations, not individuals. If the individual is the unit of a pattern, then the epidemic is the pattern itself, the system (Kranz, 1974b). The second essential of epidemiology is that it concerns not one population but two, the host and the parasite, and above all the interaction of these two populations.

Gilmour and Heslop-Harrison (1954) defined a temporary, artificial classification of plant populations which they called the deme system and which is supplementary to the more permanent, natural classification of the taxonomists. A deme is a plant population in which all individuals have a particular character in common, and the term "deme" is always used with a prefix which describes that character. A gamodeme, for example, is a population in which all individuals have a common parentage, while a topodeme is a population in which all individuals come from one locality.

Robinson (1969) proposed that, in plant pathology, the deme system be reserved for the host and that there should be a strictly parallel type system to describe parasite populations. Any one prefix can be used in both systems. A pathodeme is thus a host population in which all individuals have one pathosystem character (resistance) in common; a pathotype is a parasite population in which all individuals have one pathosystem character (parasitism) in common.

2.4.2 Differential Interaction

Provided that all other factors are constant, the amount of disease is a measure of resistance, if the pathogenicity is known; and it is a measure of pathogenicity, if the resistance is known. This is true for populations, that is pathodemes and pathotypes, as well as for individuals. Differences in the amounts of disease may be qualitative; that is, the disease is present or absent, or they may be quantitative with all degrees of difference between extremes. Throughout this discussion, the amounts of disease are measured on a scale of 0 (minimum) to 4 (maximum).

If a series of different pathodemes is inoculated with a series of different pathotypes and the amounts of disease are charted as in Figures 3 and 4, only two fundamental patterns of interaction are possible. There either is or is not a differential interaction between pathodemes and pathotypes. Figure 3 shows a differential interaction between pathodemes and pathotypes. The essential feature of this situation is that a series of pathodemes (often called differential hosts) are necessary to identify any one pathotype; and a series of pathotypes (often called physiologic or pathologic races) are necessary to identify any one pathodeme. When such a differential interaction occurs, the resistance of the pathodemes is vertical resistance and they are accordingly called vertical pathodemes. Equally, the pathogenicity of the pathotypes is vertical pathogenicity and they are accordingly called vertical pathotypes. This is the epidemiological definition of vertical resistance and pathogenicity (van der Plank, 1968). Comparison of Figures 3 and 1 shows that vertical resistance operates against some pathotypes but not others, while vertical pathogenicity operates against some pathodemes but not others.

		Vertical Pathodemes		
		A	*B*	*C*
Vertical Pathotypes	*a*	*4*	*0*	*0*
	b	*0*	*4*	*0*
	c	*0*	*0*	*4*

Fig. 3. Differential interaction

		Horizontal Pathodemes		
		D	*E*	*F*
Horizontal Pathotypes	*d*	*2*	*3*	*4*
	e	*1*	*2*	*3*
	f	*0*	*1*	*2*

Fig. 4. Constant ranking

It should be noted that a vertical pathodeme is a population in which all individuals have only a stated vertical resistance in common; they may differ in all other respects. Thus, with potato blight, vertical pathodeme R_l includes all cultivars and all wild potatoes which possess this vertical gene and no others. These cultivars and wild plants differ from each other in characters such as yield, quality, and their horizontal resistances. Similarly, vertical pathotypes, and horizontal pathodemes and pathotypes are populations in which individuals may differ in many respects other than in their stated pathosystem characteristics.

2.4.3 The Differential Interaction Fallacy

All but not only vertical pathosystems show a differential interaction between pathotypes and pathodemes. There is danger of a fallacy of the undistributed middle. We can recognize at least four kinds of differential interaction depending on the taxonomic rank of the host and parasite differential.

1. There is a differential interaction based on immunity. Thus, the various species of *Oidium* can be identified by host differentials, each of which is a different species.

2. There is a differential interaction between many species of host and the formae speciales of one parasite species, such as *Fusarium oxysporum.*

3. There is a polyphyletic differential interaction in which both the host and parasite differentials are hybrids (see Chap. 7).

4. There is a class of differential interactions in which the series of host and parasite differentials are each confined to one species. This class can be subdivided into differential interactions which either are or are not a Person differential interaction (3.2). The former constitute a vertical pathosystem; the latter do not, and may be due to an extreme loss of tolerance (5.1), or a differential interaction between resistance mechanisms and environmental factors.

2.4.4 Constant Ranking

In Figure 4, there is no differential interaction between pathodemes and pathotypes. It will be remembered that a series of *different* pathodemes were inoculated with a series of *different* pathotypes. It follows that the pathodemes can be ranked in order of resistance; thus D is more resistant than E, which is more resistant than

F. This ranking is constant, regardless of which pathotype they are tested against (otherwise there would be a differential interaction). Similarly, the ranking of the pathotypes, according to pathogenicity, is constant, regardless of which pathodeme they are tested against; d is more pathogenic than e, which is more pathogenic than f. This constancy of ranking is the only possible alternative to a differential interaction and is the definition of horizontal resistance and pathogenicity (van der Plank, 1968). D, E, and F are thus horizontal pathodemes which differ in their horizontal resistances, and d, e, and f are horizontal pathotypes which differ in their horizontal pathogenicities.

Because their ranking is constant, horizontal pathogenicity and horizontal resistance are independent of each other (van der Plank, 1968). That is, a new horizontal pathotype with an increased pathogenicity has a higher horizontal pathogenicity on all horizontal pathodemes; and a new horizontal pathodeme with increased resistance has a higher horizontal resistance to all horizontal pathotypes. Van der Plank (1968) quotes the examples of spore germination capacity and cuticle thickness respectively, where either mechanism will operate independently of the other and can be increased or decreased independently of the other.

2.4.5 Protective Chemicals

When the benomyl fungicides were used extensively, new races of pathogens appeared which were unaffected by these protective chemicals. The effectiveness of this protection thus broke down and an entirely new boom-and-bus-cycle of fungicide production became possible. For each new fungicide, there would be a matching pathogen race and there would be a differential interaction between chemicals and races. The same is true of many insecticides and other crop protection chemicals. We can reconstruct Figures 1 and 3, substituting protective chemicals for resistance genes in the host. Such chemicals and their protective mechanisms may be described as vertical.

With other chemicals, there is no differential interaction. These chemicals and their protective mechanisms may be described as horizontal. A series of different formulations of copper, dithiocarbamate and sulphur compounds would show a constant ranking in their effectiveness against all pathotypes of a given pathogen species. The same is true of a few insecticides of which pyrethrum is probably the best example, and also of rodenticides, although this pattern can be obscured by the two factors of changes in behaviour due to learning and the acquiring of antibodies.

2.5 The Histological Context

2.5.1 Capacity for Change

The histological definition of vertical and horizontal is the most precise, in pathosystem theory, because it does not lead to false syllogisms. Any resistance mechanism which is within the parasite's capacity for change confers vertical resistance;

any which is beyond confers horizontal resistance; all resistance mechanisms can thus be classified into one of these two categories.

An immediate qualification is necessary, as the phrase "capacity for change" refers to micro-evolution and not to macro-evolution. Both represent population changes due to selection pressure, but there are three important differences. Micro-evolution, as used here, involves characters which are not new, the change from one character to another is reversible, and the changes are relatively rapid, being measured on a time scale of years. Macro-evolution involves characters which are new the change is not reversible (at least in an autonomous system) and the changes are slow, being measures on a geological time scale of millions of years. This is why we say that horizontal resistance is permanent "at least in the foreseeable agricultural future". No one can be expected to postulate that horizontal resistance will endure for millions of years, even though there is good botanical evidence to suggest that it might (see p. 79).

2.5.2 *The Types of Mechanisms Involved*

The theory is easy; all resistance mechanisms can be classified into the two categories of being either within or beyond the capacity for change of the parasite. But in terms of practical experiments, this classification is not easy, nor does it coincide with any other classification of resistance mechanisms.

Let us first consider the nature of the parasite change. It involves a character which is not new, it is a reversible and relatively quick change. Which means in general that it is simple in that only one minor change is necessary in the parasite. Any mechanism which confers vertical resistance is thus likely to be a single, simple mechanism.

Conmersely mechanisms which confer horizontal resistance are likely to be either numerous, or complex, or both. Some simple resistance mechanisms are manifestly beyond the parasite's capacity for change. These include: hairiness which can confer complete resistance to both jassids and the viruses of which they are vectors; the glumes of compact barleys such as "Proctor" which are resistant to *Ustilago nuda*; the hook-like leaf hairs which puncture Heliconius larvae (Gilbert, 1971), and adhesive leaf hairs which trap aphids (Gibson, 1971). It is probable, however, that all simple mechanisms which confer horizontal resistance are physical in their action.

2.5.3 *Protective Chemicals*

Chemicals used to protect crops against parasites are essentially poisons which are either within or beyond the capacity for change of the parasite, for example Bordeaux Mixture for *P. infestans* and *Plasmopora viticola*, sulphur fungicides for powdery mildews. It seems also that the dithiocarbamate fungicides are horizontal. However, as we have seen, the benomyl fungicides are within the change capacity of a number of plant pathogens, which is true also of many insecticides and the anti-coagulant rodenticides. It should be noted, hoewever, that a protective chemical which confers vertical protection to one parasite, might confer horizontal protection to a different parasite.

Vertical protection is likely to be conferred by a single, simple mechanism. Horizontal protection is likely to be conferred by mechanisms which are either numerous, or complex, or both. For Example, Day (1974) has commented that diethyl-dithio-carbamate fungicides inhibit twenty or more enzymes and it is extremely unlikely that the physiology of a fungal parasite can change to this extent.

2.6 The Zoological Context

2.6.1 Resistance to Animal Parasites

All plants are resistant to animal parasites such as insects, mites and nematodes, their resistance being either vertical or horizontal.

Vertical resistance to insects is relatively rare. It occurs in wheat against the Hessian fly *(Mayetiola destructor)* and a gene-for-gene relationship has been demonstrated (Day, 1974). It occurs rather more commonly against nematodes.

Horizontal resistance to insects is common, occurring both with postharvest and preharvest pests. Thus, rice grains with well-formed husks are resistant to at least four species of storage pests when compared with grains which have inherited husk defects (Link and Rossetto, 1972). However, except when there has been a prominent, single resistance mechanism, horizontal resistance to insects has been largely ignored by plant breeders. This is perhaps encouraging as the law of diminishing returns has not yet begun to operate, and great advances are thus waiting to be made. The same is true of horizontal resistance to other animal pests such as nematodes and mites.

2.7 The Pathosystem Context

2.7.1 Two Categories of Interaction

There are only two categories of host-parasite interaction, differential interaction and constant ranking. Any interaction, regardless of its agricultural effects, mechanisms or inheritance, must belong to one of these two categories.

2.7.2 Two Categories of Infection

An epidemic is a system which is divisible into sub-systems and we must now discuss two sub-systems of special relevance to pathosystem analysis. In this discussion, coffee leaf rust, caused by *Hemileia vastatrix*, is used as an example but, with only minor qualifications, the principles involved apply to all plant disease epidemics. Coffee rust is a good example because coffee is a tree with a large population of individual leaves and the rust has two kinds of spore dispersal.

Nutman and Roberts (1963) showed that the spores of *H. vastatrix* are water-dispersed in Kenya coffee plantations. More recently, the spread of coffee rust in South America shows that wind-dispersal also occurs. The mechanism of this dual function of the rust spores is not understood and some of the details appear to be in conflict. Nevertheless, we can assume both methods of dispersal for the purposes of pathosystem analysis.

The main function of water-dispersal is to carry the pathogen from leaf to leaf within an individual coffee tree; the main function of wind-dispersal is to carry the pathogen from one coffee tree to another, particularly in a relatively sparse population of wild coffee trees growing in mixed forest. We can thus divide the coffee rust epidemic into two sub-systems on the basis of two kinds of infection. With water-dispersal, the host is normally infected from its own lesions; only one individual tree being involved; this is called auto-infection. With wind-dispersal, the host is infected from the lesions of a different host tree; two host individuals being involved, the infected and the infector, the recipient and the donor; this is called allo-infection.

These two sub-systems assume such importance in pathosystem analysis that they justify new terms.

2.7.3 The Esodemic and the Exodemic

The sub-system which involves auto-infection only may be called the esodemic and the sub-system which involves allo-infection only may be called the exodemic.

There are also two kinds of resistance; vertical and horizontal. It will transpire that vertical resistance can only reduce the exodemic and that the esodemic can only be reduced by horizontal resistance. (However, there is again a danger of the fallacy of the undistributed middle; horizontal resistance can also reduce the exodemic. It follows that not only vertical resistance reduces the exodemic and not only the esodemic is reduced by horizontal resistance.)

We may thus define the epidemic as the spread of a pathogen population within a host population; the exodemic is the spread of a pathogen from one host individual to another by allo-infection; and the esodemic is the spread of a pathogen population within one host individual by auto-infection.

As precision in biological definitions is invariably difficult, some qualification is inevitable.

The esodemic involves host individuals; each individual has its own esodemic. But, because the population consists of individuals, each of which eventually becomes diseased, the esodemic has a population effect and is an essential component of the epidemic. The esodemic involves auto-infection only. Coffee rust is a good example because coffee is a tree with a large population of individual leaves, but the principle remains true for any host individual, however small. If a disease has sporulating lesions, as with coffee rust, the esodemic increases by auto-infection; if it is a systemic disease, the esodemic increases without auto-infection. But whatever the nature of the disease, the essential feature of the esodemic is that there is no allo-infection. The esodemic involves host tissues which all belong to one host individual and which consequently all possess the same vertical susceptibility.

The exodemic involves the entire host population but is only the spread of disease from one host individual to another, involving only allo-infection. However, if the pathogen has a wide host range, or is a heteroecious rust, the infector host may belong to a different species from the infected host. Similarly, if the pathogen is a facultative parasite, allo-infection may derive from a dead substrate. In no instance, however, does the exodemic involve auto-infection. The essential feature of the exodemic is that it involves two lots of host tissue which are not necessarily identical with respect to vertical resistance. This means that allo-infection may involve a spore whose vertical pathogenicity does not match the vertical resistance of the host; the vertical resistance is then effective.

In systems analysis, it is more important to distinguish between differences of pattern than between differences of level of pattern. The exodemic involves a population, the esodemic an individual. The exodemic involves the pattern, the esodemic the units of that pattern. Wild coffee is a population of individual trees, whereas a cultivated coffee crop may be a single clone; it is then the epidemiological equivalent of a single coffee tree, every coffee plant in the crop being genetically identical, just as every coffee leaf in one tree is genetically identical. The patterns are the same but they occur at different system levels. In the wild coffee rust pathosystem we thus refer to a population of individual trees; in the cultivated coffee pathosystem we refer to the population of individual clones. The terms esodemic and exodemic are systems terms and were defined as such. They define the pattern, not the level pattern. When using the terms esodemic and exodemic, therefore, it is essential to state at which level of pathosystem they are employed. It may prove convenient to do this by qualifying them with a definition of the individual involved; thus the plant-esodemic and the crop-esodemic; the plant-exodemic and the crop-exodemic.

2.8 The Esodemic

The esodemic involves continuity of host tissue, being the spread of disease within the components (e.g. leaves, etc.) of one host individual. If that host individual possesses vertical resistance, all its components possess the same vertical resistance. The following seven conclusions can be drawn.

2.8.1 The Breakdown of Vertical Resistance

If a coffee tree has no rust whatever, it can only be infected by a spore coming from another host individual, that is by allo-infection. If the vertical pathogenicity of that spore does not match the vertical resistance of the coffee tree, the allo-infection will fail. It follows that the esodemic can only begin with a matching allo-infection; in other words, the esodemic begins with the breakdown of vertical resistance.

2.8.2 Vertical Resistance Valueless in the Esodemic

Within an individual coffee tree, each leaf has the same vertical resistance. If rust is present in the tree, it is a vertical pathotype which matches the vertical resistance of that tree. Each rust pustule will produce spores which match the vertical resistance of all the leaves of the tree. It is self-evident that vertical resistance is valueless against auto-infection and that, consequently, it has no function whatever in the esodemic.

2.8.3 Horizontal Resistance Is Essential in the Esodemic

Because vertical resistance is valueless in the esodemic, the only defence against auto-infection is horizontal resistance. This is possibly the best argument for the postulation (5.2) that horizontal resistance occurs in all plants against all parasites, even if it is agriculturally inadequate in many cultivars. It is clear also that horizontal resistance is the resistance which invariably remains after vertical resistance has broken down, however inadequate its level may be in many cultivars. If a coffee tree possessed no horizontal resistance whatever, the rust esodemic would increase without hindrance and the tree would die from loss of leaf. In Ethiopia, where *Coffea arabica* can survive in a wild state, rust is ubiquitous, but is not a serious disease. The level of horizontal resistance is clearly adequate to control the esodemic in every tree, a situation pertaining in all wild host populations which would otherwise become extinct.

2.8.4 Vertical Resistance Is Temporary in Uniform Crops

Crop uniformity is an essential feature of agriculture and often results in all the individual plants of a cultivated crop possessing the same vertical resistance. That crop is then the epidemiological equivalent of a single coffee tree, and each individual plant within the crop the equivalent of a single leaf within the coffee tree. There is only one vertical resistance for the crop as well as for the tree, and it breaks down with the first successful allo-infection of one plant, just as the vertical resistance of a coffee tree breaks down with the first successful allo-infection of one leaf. The esodemic then begins, and, as we have seen, vertical resistance is valueless against auto-infection.

When a single clone of potatoes or a single pure line of wheat is protected by vertical resistance, that protection will last only until a successful allo-infection occurs. The crop-esodemic then begins and the vertical resistance has no further function. Thus vertical resistance is temporary resistance in uniform crops and this is the basis of the boom-and-bust cycle of cultivar production.

In practice, there is great variation in the time required for a crop breakdown of vertical resistance. In some diseases, such as *Puccinia polysora* of maize, the breakdown occurs so quickly that there is no time for a vertical pathodeme to be multiplied as a new cultivar. In others, such as cabbage yellows (*F. oxysporum* f. sp. *conglutinans*) the breakdown is so delayed that the vertical resistance is all but permanent. It is the middle time scales, occuring typically in the small-grain cereal rusts which are dangerous. The vertical resistance endures long enough for a boom in a new cultivar, but not long enough to prevent a bust at about the time when the boom is reaching its maximum.

2.8.5 The Measurement of Horizontal Resistance

We can only measure the level of horizontal resistance after the esodemic has started. In other words, we must use a matching vertical pathotype and this, in its turn, means that it is impossible to measure the horizontal resistance of a cultivar whose vertical resistance has not yet broken down. This simple conclusion has been consistently overlooked during some 70 years of breeding plants for resistance. The distinction between the esodemic and the exodemic is of crucial importance in pathosystem management, and the failure to appreciate this has led to horizontal resistance remaining unrecognized and unused.

2.8.6 Breeding for Horizontal Resistance

If we wish to screen plants for horizontal resistance, we can only do so under conditions in which that resistance is operating, that is, we must screen for horizontal resistance during the esodemic, or after vertical resistance has broken down. Once again, this simple conclusion has been consistently overlooked during more than half a century of plant breeding for resistance. Plants have been screened against non-parasitic populations of the parasite (4.3) and incredible though this may seem, it explains why there has been not only a failure to increase horizontal resistance but even an actual loss of horizontal resistance due to the vertifolia effect (4.3). It also explains why composite crosses in wheat and barley failed to accumulate horizontal resistance.

2.8.7 The Recovery of Vertical Resistance

If a coffee tree were deciduous, the esodemic would end with the completion of leaf fall; later the new leaves would be rust-free and their vertical resistance would have recovered in that it would again provide useful protection against allo-infection. Recovery is the converse of breakdown, it concerns the effectiveness of the resistance which remains unaltered in the face of a changing pathogen population. The esodemic thus begins with the breakdown of vertical resistance and ends with its recovery, a conclusion of major importance in pathosystem analysis and discussed in detail later (3.3).

Although coffee is not deciduous, it grows in areas where there is a marked dry season each year, when infection cannot occur, and a natural defence mechanism ensures that rusted leaves are prematurely shed. In a natural pathosystem, therefore, the esodemic ends with each dry season. An alternate host has yet to be identified but, if it occurs, it presumably carries the sexual phase of the rust and produces a wide range of different vertical pathotypes at the onset of new rains. In the artificial, or crop, coffee pathosystem, this delicate balance is often lost, coffee being cultivated as a pure line and with a high host population density. The effects of the exodemic are lost and the horizontal resistance is often inadequate to control the esodemic, subsequent defoliation often being excessive. An unsuitable climate, irrigation or an ill-timed application of fungicide can both intensify the esodemic and make it continuous.

2.9 The Exodemic

The exodemic involves discontinuity of host tissue; it is the spread of disease from one host individual to another by allo-infection. The infector host may have a different vertical resistance from the infected host, and will then produce spores with a vertical pathogenicity which does not match the vertical resistance of the infected host. The following seven conclusions can be drawn.

2.9.1 The Sole Function of Vertical Resistance

Except for the very rare mutation, the vertical pathogenicity of the spores of a coffee rust pustule matches the vertical resistance of the host coffee tree. If those spores are then carried by wind to another tree with a different vertical resistance, the allo-infection will fail; this is the sole function of vertical resistance in nature. Vertical resistance can only prevent infection and it can only prevent allo-infection. Because some matching inevitably occurs, vertical resistance cannot prevent all allo-infection, it can only reduce the exodemic.

Some qualification is necessary. The term infection can be used in two different contexts which must be clearly distinguished. In a histological context, infection means the penetration of the host; once penetration is complete, colonisation begins. Vertical resistance, as manifested by the hypersensitive reaction, normally operates after infection is complete and when colonisation has already begun. In this context, therefore, it would be incorrect to say that vertical resistance prevents infection; it prevents colonisation. But the term infection can also be used in an epidemiological context; thus auto-infection and allo-infection and, in this context, it is legitimate to say that vertical resistance prevents infection.

A second qualification concerns incomplete vertical resistance which is discussed later (3.1). We have seen that vertical resistance can only reduce the frequency of matching allo-infection but incomplete vertical resistance can do even less. If a host with incomplete vertical resistance is infected by a matching vertical pathotype, the vertical resistance fails to operate. If that same host is infected by a non-matching vertical pathotype, the infection does not fail, the colonisation of the host is merely reduced. It would be misleading to compare this situation with the reduction of the esodemic by horizontal resistance which occurs after the breakdown of vertical resistance. Incomplete vertical resistance confers incomplete protection before its breakdown and no protection whatever after its breakdown.

2.9.2 Vertical Resistance Is Permanent in Nature

Consider a population of wild coffee plants which is so genetically mixed that, within a given area, every individual tree has a different vertical resistance. No individual tree can allo-infect a near neighbor or be allo-infected by it. Epidemiologically, this pattern of vertical resistances is the equivalent of a reduction in both the host and the pathogen population densities. The host population density is reduced to that of any one vertical pathodeme, while the pathogen population density is reduced to that of any one vertical pathotype, and the number of successful allo-infections is greatly reduced. In such a pattern, a single breakdown of vertical resistance occurs only in individuals and never in the population as a whole. Even if every individual is successfully allo-infected, many different breakdowns of vertical resistance are involved. Such a process requires time, and will always require time so long as the host population heterogeneity is maintained.

Such a pattern of different vertical resistances has two properties. The epidemic is slowed down and this effect is permanent; these are the epidemiological properties of horizontal resistance. A pattern of different vertical resistances thus produces an *apparent* horizontal resistance (3.2). In practice, the pattern is somewhat more complicated than presented here, due to the possibility of overlapping vertical genomes (3.2) but the general principle is unaffected.

In agriculture, a comparable effect can be produced with a multiline (Jensen, 1952), that is, a host population in which there is a mixture of different vertical pathodemes which are otherwise closely similar to each other. There has, perhaps, been too much caution in the production of agricultural multilines, due mainly to fears about the so-called "super-race" of the pathogen, a complex vertical pathotype which possesses all the relevant vertical genes and which can accordingly match every vertical resistance in the multilines; but it is also a myth. There is evolutionary evidence against the super-race as it would prevent the evolution of vertical resistance. Vertical resistance reduces population densities; if this mechanism did not work in the wild pathosystem, vertical resistance would not evolve. Suppose there were five vertical genes in one complex, vertical pathodeme which is cultivated as a clone. The matching, complex, vertical pathotype would then have a high survival value and the selection pressure in its favour would be great. But, if each vertical gene occurred singly in a multiline of five monogenic, vertical pathodemes, the selection pressure for the complex vertical pathotype with all five genes would not occur. Whichever host individual it infected, it would possess four unnecessary vertical genes which would tend to be lost due to negative selection pressure. Survival values tend to balance; one is gained at the expense of another, but unnecessary survival values are not values at all; they are hindrances (5.5).

2.9.3 The Value of Vertical Resistance in Agriculture

As we have seen, vertical resistance can only reduce the frequency of allo-infection. If this effect is to be marked, there must be a discontinuity of host tissue. This point is elaborated at 3.3 and a distinction is made between spatial and sequential discontinuity. As a general rule, the greater the discontinuity, the more valuable is vertical resistance. (But this is not the only factor; see 4.3–4.5.) All agricultural development, however, has tended towards continuity of host tissue. Factors such as bulk production, storage, distribution and processing all demand uniformity of product. This is one of the many ways in which agricultural plant pathosystems differ from wild plant pathosystems. We recognize the necessity for this uniformity but we must also accept the consequences. The main consequence is a considerable reduction in the usefulness of vertical resistance in agriculture.

2.9.4 The Relative Values of Vertical and Horizontal Resistance in Nature

Vertical resistance can do no more than reduce allo-infection and slow down the exodemic. Horizontal resistance can also do this, and can in addition reduce auto-infection and all subsequent stages of the esodemic.It is clear that all plants which possess vertical resistance also possess horizontal resistance; many plants possess horizontal resistance but no vertical resistance (6.6), but no plant possesses vertical resistance only. Such a plant would be totally destroyed by the esodemic. In nature, therefore, horizontal resistance is essential and universal, while vertical resistance is only a supplementary protection which occurs in certain heterogeneous host populations.

2.9.5 The Progenitors of Crop Species

If a crop species, such as wheat or potatoes (Solanum), possesses vertical resistance to one or more of its pathogens, the progenitors of that species must have grown in populations which were heterogeneous with regard to vertical resistance because vertical resistance has no survival value and, consequently, would not evolve in a genetically uniform population (3.3). Even in an inbreeding population, there is always a minimum of cross-pollination which will maintain both population heterogeneity and the value of vertical resistance.

If we can show that its wild progenitors grew (or still grow) in genetically uniform populations, we can be confident that no vertical resistance will occur in a crop species. The converse that if a crop species has no vertical resistance to any of its pathogens, its progenitors probably grew in uniform populations, is probably but not necessarily true, because vertical resistance could fail to evolve in a genetically mixed population. It seems, for example, that vertical resistance to *P. polysora* in maize is rare and of minor epidemiological significance.

There is danger of another fallacy of the undistributed middle. All vertically resistant crops had mixed progenitor populations, but not only vertically resistant crops did so. Uniform progenitor populations produced crops without vertical resistance but not only uniform progenitor populations did so.

2.9.6 Population Properties of the Host

If vertical resistance is to have survival value, allo-infection and the exodemic must be a major component of the epidemic, which necessitates both spatial discontinuity of host tissue, as in heterogeneous populations, and sequential discontinuity of host-tissue, as in annuals or the leaf diseases of a deciduous tree. Therefore vertical resistance is likely to be more common in annuals than in perennials; in deciduous than in evergreen perennials; in temperate plants, with their marked dormancy, than in tropical plants; and in leaf diseases than in root diseases because leaves can be shed while roots cannot. This topic is discussed in greater detail at p. 44.

2.9.7 The Value of Horizontal Resistance in Vertically Resistant Crops

If vertical resistance occurs in a crop species, the progenitors of that species grew in mixed populations with many different vertical resistances. The effect of such a pattern is that of an apparent horizontal resistance. If the influence of this apparent horizontal resistance was great in the progenitor population, the need for true horizontal resistance would have been reduced, in which case natural levels of horizontal resistance in the progenitors may then be inadequate to control a disease under agricultural conditions of population uniformity (9.2).

Chapter 3 Vertical Pathosystem Analysis

3.1 General

3.1.1 Recapitulation

The term vertical is derived from the diagram in Figure 1. It is an abstract term which, having no literal or descriptive connotations, can be defined and used with precision in widely different contexts.

Agriculturally, vertical resistance is usually temporary because its effectiveness is liable to break down due to a change in the parasite population. In spite of exceptions, it usually provides a complete but impermanent control of a pest or disease. Its use thus leads to a boom-and-bust cycle of cultivar production.

Vertical resistance is probably always inherited oligogenically but not all oligogenic resistance is vertical. Vertical resistance may provide either a complete or an incomplete protection and its effects are then qualitative or quantitative respectively.

Vertical resistance and parasitism are indicated by a differential interaction in the amounts of disease when a series of different pathodemes is inoculated with a series of different pathotypes, but not all differential interactions are due to vertical resistance.

Vertical resistance is conferred by mechanisms which are within the capacity for micro-evolutionary change of the parasite. Such mechanisms are likely to occur singly and to be simple.

The esodemic involves auto-infection and vertical resistance is valueless against it; the esodemic begins with the breakdown of vertical resistance and ends with the recovery of vertical resistance. The crop esodemic is the same pattern as the plant esodemic but occurs at a higher pathosystem level, thus vertical resistance is temporary in uniform crops.

The exodemic involves allo-infection. The sole function of vertical resistance is to reduce allo-infection and, hence, to reduce the exodemic; an effect permanent so long as host population heterogeneity is maintained. The need for crop uniformity limits the value of vertical resistance in agriculture. In a natural pathosystem, vertical resistance, if it occurs, is supplementary to horizontal resistance. Vertical resistance occurs in crop species whose progenitors grew in heterogeneous populations and cannot evolve in uniform populations. It has survival value only in host populations which exhibit a marked spatial and sequential discontinuity of host tissue.

3.1.2 Incomplete Vertical Resistance

Figure 1 concerns potato blight caused by *P. infestans*. This vertical resistance confers complete protection against non-matching vertical pathotypes, apart

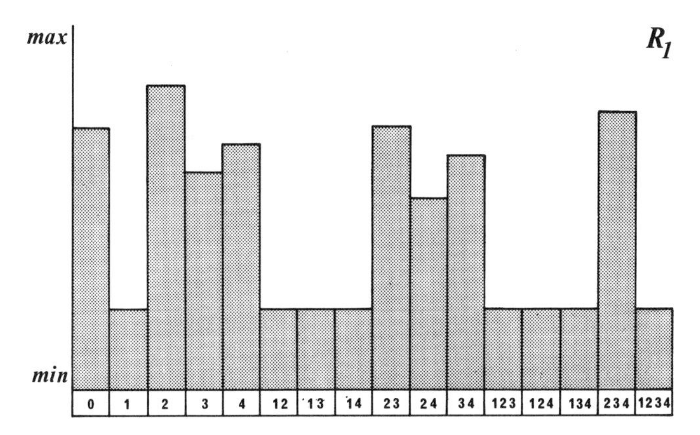

Fig. 5. Incomplete vertical resistance

from a few hypersensitive flecks which are almost microscopic and without epidemiological significance.

With other diseases, notably some of the cereal rusts, the vertical resistance mechanism often provides incomplete protection against non-matching vertical pathotypes. Figure 1 would then become Figure 5 and the zeros of Figure 3 could be any value below 4. The term "incomplete vertical resistance" is used to describe this phenomenon.

Incomplete vertical resistance can be confusing, and can be mistaken for horizontal resistance because its effects are quantitative. There are thus two kinds of quantitative resistance. However, quantitative vertical resistance is inherited oligogenically and quantitative horizontal resistance polygenically.

Incomplete vertical resistance is also a nuisance, having all the disadvantages of complete vertical resistance with additional drawbacks of its own. It does not provide a complete disease control, its use increases the flexibility of the pathogen population dynamics (3.5); and it is responsible for many difficulties when working with horizontal resistance. It could also be dangerous if it were genuinely mistaken for horizontal resistance in a breeding programme.

Histologically, an incomplete vertical resistance is apparently due to a time factor, where the resistance mechanism operates, but too slowly. Host colonisation and pathogen reproduction can then occur but only at a reduced rate. However, other explanations may be found.

It is perhaps worth enquiring why incomplete vertical resistance should occur at all, as its survival value in a natural pathosystem is obviously limited. One possible explanation is that it is an artifact of the crop pathosystem. It is possible, for example, that the mechanism is operating in an unsuitable environment and that, were it to operate in the optimum environment of the wild progenitors, it would confer complete protection.

3.1.3 Apparent Vertical Resistance

There are a number of situations in which it is possible to conclude, falsely, that we may be working with vertical resistance. The most obvious of these is when there is an apparent breakdown of resistance, such as may occur, for example, in the compact barleys in which the stigma is protected from infection by *U. nuda hordei*. In certain weather conditions, the flowers may open and this oligogenic, horizontal resistance mechanism then fails to operate. This is a physiological breakdown; it is not due to a change in the pathogen population and does not lead to a boom-and-bust cycle. Similarly, inadequate disease testing can lead to the appearance of high resistance which proves illusory only at a later date. There have been occasional resurgences of an old disease such as sugarcane mosaic for this reason.

The differential interaction fallacy is a major source of confusion. Resistance to sugarcane smut *(Ustilago scitaminea)* was long believed to be vertical because cultivars resistant in one region were susceptible in another, and it was concluded that different vertical pathotypes occurred in various parts of the world. Thus, Co. 421 was resistant in India but susceptible in Kenya. Later, it was shown that the Kenya stocks of this cultivar were heavily contaminated with the susceptible Co. 419 and that, when purified, Co. 421 was as resistant in Kenya as in India. Waller (1970) showed that the level of resistance to this disease was correlated with bud morphology and that the latter character varied with environment. There can thus be a differential interaction between horizontal pathodemes and environmental factors.

It must also be remembered that it was fashionable to believe that all resistance would breakdown sooner or later. Some workers seemed to want such breakdowns; they prophesied it and longed to have their prophesies proved correct, but they were often mistaken and their mistakes were usually due to the fallacy of the undistributed middle, of which we have already noted some examples. Not all oligogenic resistance is vertical; not all qualitative resistance is vertical. Not all differential interactions are due to vertical resistance (see Chapt. 7), not all simple and single mechanisms confer vertical resistance; not only vertical resistance reduces the exodemic; and not only the esodemic is reduced by horizontal resistance.

We can note others; not all vertical resistance is due to hypersensitivity; the vertical resistances to Fusarium and Verticillium wilts, and to loose smut *(U. nuda)* of barley are due to quite different mechanisms; not all hypersensitivity confers vertical resistance; Müller (1950) induced hypersensitive reactions in crops such as onion, lettuce and dahlia with *P. infestans*; not all breakdowns are due to vertical resistance (5.5); not all horizontal resistance is permanent (5.5), and not only horizontal resistance is permanent, particularly if there is sound pathosystem management.

At this point, it is perhaps instructive to note in which contexts this fallacy cannot arise. Apparently there are three. (1) *All* resistance mechanisms which are within the parasite's capacity for change confer vertical resistance and *all* vertical resistance is due to such mechanisms. (2) It is probably true that *all* gene-for-gene relationships (3.2) result in vertical resistance and *vice versa*. (3) It is likely that *all* vertical pathosystems and *only* vertical pathosystems show a Person differential interaction (3.2).

This discussion may seem like hair-splitting, but is not. It is essential for three reasons. (1) We cannot hope to achieve effective pathosystem management if our pathosystem analysis is faulty. (2) The discussion emphasises once again the necessity for abstract terms such as vertical and horizontal which can be differently but accurately defined in each of many contexts. Those definitions can also be "sharpened" as the pathosystem concept develops. (3) A scientific paper can be reduced to stark nonsense by the presence of even one false syllogism. This is a double warning, both to those intending to write about this subject, and to those intending to read about it. The existing scientific literature is sadly imperfect in this respect.

3.1.4 *Immunity*

Immunity is an absolute quality; it is a non-variable survival value. A plant either is or it is not immune to a parasite, and anything less than immunity is resistance which is a variable survival value. Thus, coffee is immune to wheat rust but is merely resistant to coffee rust, however high the level of that resistance may be.

The absolute quality of immunity means two things. Immunity confers protection which is both complete and permanent. Resistance may confer a complete protection or a permanent protection but it never confers both. Thus, vertical resistance can confer complete protection to some but not all vertical pathotypes; it is then temporary resistance which delays the onset of the esodemic. Conversely, in a heterogeneous host population, the effects of vertical resistance on the exodemic are permanent; but the control of the exodemic is incomplete.

It is possible to equate some (but not all) of the characteristics of vertical resistance with immunity. Van der Plank (1975) comments that some features of vertical resistance, such as a differential interaction, and some aspects of its mechanisms and inheritance, apply also to immunity. So also do some of the features of frozen vertical resistance (4.5). However, this discussion is beyond the scope of the present book and interested readers should consult the original. Comparable arguments can also be made for equating absolute horizontal resistance (5.1) with immunity.

Perhaps the best view is that immunity is not part of the pathosystem concept; it is outside the conceptual boundaries of the pathosystem. When analysing and managing a potato pathosystem, we normally ignore the possibility of wheat rust spores being present in the potato crop.

3.2 The Gene-for-Gene Relationship

3.2.1 *Definitions*

The gene-for-gene relationship was discovered by Flor (1942, 1959) and was formally defined by Person *et al.* (1962). It means that, for every vertical gene in the host population, there is a corresponding, matching vertical gene in the parasite population. Vertical resistance is effective if the vertical parasitism genes do not match the vertical resistance genes, but fails to operate when the vertical genes do match. Person (Personal communication 1973) comments that the most succinct definition is that "the demonstration of either depends on the presence of both". That is, a vertical resistance gene can only be identified with the matching vertical parasitic gene, and *vice versa*. We can extend this argument to populations and vertical genomes. The identification of a vertical pathodeme, which may consist of many different cultivars, depends on the matching vertical pathotype, and the identification of a vertical pathotype, which may consist of many different horizontal pathotypes, depends on the matching vertical pathodeme.

Flor's (1942, 1959) discovery stimulated a large amount of research. As a result, it is probably safe for us now to conclude that there is no danger of fallacies of the undistributed middle. All vertical pathosystems involve a gene-for-gene relationship; and only vertical pathosystems do so.

3.2.2 The Nature of the Matching

The gene-for-gene relationship has been called the matching gene theory. Strictly, it is not the genes which match, but the characters, whose inheritance is controlled by those genes. At the molecular level of the pathosystem, we can think of vertical resistance and pathogenicity in terms of chemical pathways which either do or do not match each other. No doubt, the details of the chemistry are complex, but the principle and the pathosystem effects are simple.

3.2.3 Terminology

A vertical genome may contain many, few, one or even no vertical genes. If it possesses one or only a few vertical genes, it is called simple; if it possesses many vertical genes, it is called complex. We can also refer to simple and complex vertical resistances, pathodemes, pathogenicities, parasitic abilities and pathotypes. The pathodeme with no vertical genes has been called the universal susceptible. In spite of its lack of vertical genes, it is a vertical pathodeme because of its vertical susceptibility, and is a differential host. Similarly, a pathotype with no vertical genes is a vertical pathotype because it is used in the indentification of vertical pathodemes.

We must define the term "matching". If a pathodeme has more vertical genes than a pathotype, or has an equal number of different vertical genes, the vertical genomes do not match and the resistance is effective. If the pathodeme and pathotype have identical vertical genomes, they match and the resistance fails. If the pathotype has all the vertical genes of a pathodeme, and some others in addition, the resistance will again fail and the vertical pathotype and pathogenicity are described as overlapping. One vertical resistance can thus be overcome by a number of overlapping, vertical pathotypes (Fig. 1). Overlapping may be more common than we realise because an apparently matching vertical pathotype may possess vertical genes which have not yet been recognised. However, when no confusion is likely to arise, overlapping vertical pathotypes and pathogenicities can be described as matching for the purposes of general description.

The possible numbers of different vertical genomes in 2^n, where n is the number of vertical genes, this number including the genome with no vertical genes. It will be seen shortly that this simple mathematical relationship is useful when it comes to the naming of vertical genomes.

It is clear that the number of different vertical genomes can be very large. With 6 genes, there are 64 possible genomes; with 35 genes, there are 34, 359, 738, 368 possible genomes. Some of the earlier workers laid great stress of the large number of pathogen "races" which had been identified, but as more than 1,000 vertical genomes are possible with only 10 vertical genes, it is clearly more useful to identify the genes than the genomes.

Vertical genes and genomes are identified by matching in the host and parasite; the demonstration of either requires the presence of both. A series of host differentials (vertical pathodemes) is used to identify any one vertical pathotype, and a series of parasite differentials (vertical pathotypes) is used to identify any one vertical pathodeme. Ideally, the differentials in the host series should each possess only one vertical gene and the parasite differentials should each lack only one vertical gene. Each series should also include the differential with no vertical genes. The number of differentials is then at its minimum and the identification of any genome at its most simple.

3.2.4 Nomenclature

Terms have a linguistic meaning and can be employed in a system of logic. The less precise their meaning, the less useful they are. Names do not require a linguistic meaning, being only labels. This is why an imprecise word such as "race" is acceptable as a name (e.g. Race 1234 of *P. infestans*), but not as a term.

Names are used to identify the units of a pattern, at various systems levels, in particular, to label the various ranks of a taxonomic hierarchy. A name can thus be given to an individual organism, a population within a species, a species within a genus and so on. The identification must be accurate but the only important criterion of a name is that it must be unique within the system. If two ranks have the same name, they cannot be conceptually distinguished.

In discussing the naming of vertical genes, genomes, pathotypes and pathodemes, we shall take coffee leaf rust as our example, referring to other pathosystems as necessary.

Mayne (1932, 1935, 1936) first discovered "physiologic specialisation" in coffee rust in India and identified four vertical pathotypes which he named Races 1–4. Later the Coffee Rust Research Centre was set up in Portugal where coffee is not a commercial crop and isolates of the pathogen could be imported without danger. This Centre, under the direction of Branquinho d'Oliveira, was eminently successful in elucidating the vertical pathosystem of coffee rust.

Initially, the vertical genes were unidentified, only the two series of differentials being used to identify each other. The host differentials were labelled with letters and the pathogen differentials with Roman numerals, both in chronological order of discovery (Fig. 6). Later, Bettencourt and Noronha-Wagner (1971) identified five vertical genes.

One of the benefits of the gene-for-gene relationship is that matching vertical genes and genomes can be given the same name without necessarily indicating whether they occur in the host or the parasite. Matching vertical pathotypes and pathodemes can then be given a common designation which also indicates which vertical genes are present.

Old Designations		New Designations		
PATHOTYPES	PATHODEMES	BLACK	HABGOOD	
		GENES / NAME	GENES	NAME
IV	β	0		0
	α	1	2^0	1
		2	2^1	2
		1 2	2^0 2^1	3
		3	2^2	4
		1 3	2^0 2^2	5
		2 3	2^1 2^2	6
		1 2 3	2^0 2^1 2^2	7
	y	4	2^3	8
XIX	I	1 4	2^0 2^3	9
		2 4	2^1 2^3	10
		1 2 4	2^0 2^1 2^3	11
		3 4	2^2 2^3	12
		1 3 4	2^0 2^2 2^3	13
		2 3 4	2^1 2^2 2^3	14
		1 2 3 4	2^0 2^1 2^2 2^3	15
II (I)[+]	E	5	2^4	16
III	C	1 5	2^0 2^4	17
I (II)[+]	D	2 5	2^1 2^4	18
XVII	L	1 2 5	2^0 2^1 2^4	19
VII (III)[+]	G	3 5	2^2 2^4	20
		1 3 5	2^0 2^2 2^4	21
VIII (IV)[+]	H	2 3 5	2^1 2^2 2^4	22
XII	V, Z	1 2 3 5	2^0 2^1 2^2 2^4	23
XV	J	4 5	2^3 2^4	24
X	W	1 4 5	2^0 2^3 2^4	25
XXIV	O	2 4 5	2^1 2^3 2^4	26
XXIII		1 2 4 5	2^0 2^1 2^3 2^4	27
		3 4 5	2^2 2^3 2^4	28
		1 3 4 5	2^0 2^2 2^3 2^4	29
XIV	T, X	2 3 4 5	2^1 2^2 2^3 2^4	30
XVI	S, U	1 2 3 4 5	2^0 2^1 2^2 2^3 2^4	31

([+] Mayne's (1936) designations)

Fig. 6. Vertical gene nomenclature (Coffee leaf rust)

Black *et al.* (1953) proposed such a nomenclature for potato blight caused by *P. infestans*. Each vertical gene was given an Arabic numeral in chronological order of discovery. The numerals of one genome are enclosed in brackets and when more than nine vertical genes are known the numerals are separated by commas e.g. (1, 2, 12). Complex genomes can be abbreviated with a hyphen representing all intermediate numerals in a continuous series, e.g. (1–5, 7–10). Matching vertical pathotypes and pathodemes thus have the same name which also indicates the composition of the genome.

Habgood (1970) proposed an ingenious nomenclature which makes use of the fact that the number of possible genomes is 2^n, where n = the number of vertical genes. Robinson (1974) applied Habgood's nomenclature to the vertical genes of coffee rust (Fig. 6). Each vertical gene is named in chronological order of discovery, $2^0, 2^1, 2^2, 2^3$ etc. These numerals have the values 1,2,4,8,16 etc, each value being double that of its predecessor. This has the advantage that the sum of any one combination of values is unique within the system. Thus, a genome with the first five genes would have a sum of 31 (i.e. $1+2+4+8+16$) and no other combination of genes has this sum. It is then possible to refer to Vertical Pathotype 31 and Vertical Pathodeme 31 which clearly match each other and whose vertical genes are readily identifiable. The genes are indentified by subtraction of the highest possible gene value from the sum and from each remaining sum. Thus $31-16=15$; $15-8=7$; $7-4=3$; $3-2=1$ and we know that genes 16, 8, 4, 2, and 1 are present in the genome.

Several comments are necessary. Habgood (1970) was concerned with disease in which the vertical genes were not clearly identified and in which there may be other complicating factors such as modifier genes and incomplete vertical resistance. He consequently found it convenient to apply his nomenclature to the host differentials rather than to the vertical genes. This application is eminently practical and has been successfully adopted by Johnson *et al.* (1972) in the *Puccinia striiformis* wheat pathosystem. Nevertheless, the ultimate aim should always be the identification and naming of the vertical genes.

The problem of incomplete vertical resistance was solved by Habgood (1970) by fixing a point on the scale of disease above which any level of disease rates as susceptible and any level below as resistant. Obviously, indeterminate results are possible and lead to confusion, but incomplete vertical resistance is confusing in any event and this system reduces the confusion to the minimum.

Finally, we should note a symmetrical pattern in the Habgood nomenclature (Fig. 6). It will be observed that when the vertical genomes are arranged in the numerical order of their sums (Habgood names), the genes with value one occur at every alternate unit; the genes with value two occur at every alternate pair; the genes with value four occur at every alternate foursome; and so on to infinity. This symmetry becomes significant in the Person/Habgood differential interaction (Fig. 8).

3.2.5 Demonstration

The demonstration of a gene-for-gene relationship is both the most reliable and the most practical way of demonstrating the vertical nature of a pathosystem. At first sight, such a demonstration is difficult because it involves genetical studies in both the host and the parasite combined with related interaction studies. Apart from its inherent difficulties, this demonstration is often impossible because the sexual stage of the parasite is either absent or unknown. For example, functional oospores of *P. infestans* do not occur outside Mexico and the sexual phase of coffee leaf rust *(H. vastatrix)* has yet to be discovered. An alternative and easier demonstration would thus be very valuable.

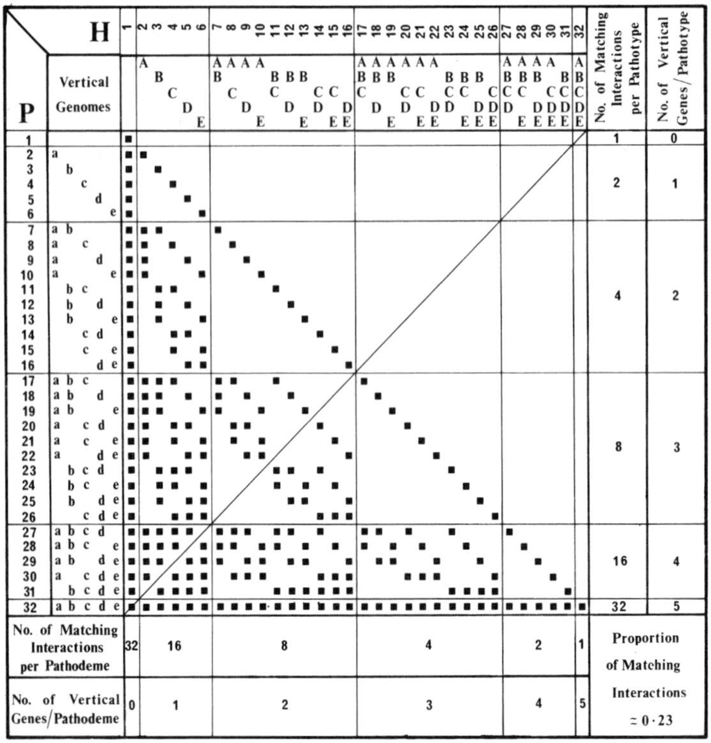

Fig. 7. The Person differential interaction

Person (1959) has provided such an alternative in a paper which has been sadly neglected. He demonstrated the mathematical properties of a theoretical parasitic system involving a gene-for-gene relationship. If the same mathematical properties can be demonstrated for a real parasitic system, then that system also involves a gene-for-gene relationship. This demonstration is a special category of differential interaction which may be called the Person differential interaction.

3.2.6 *The Person Differential Interaction*

Figure 7 shows the differential interaction between the thirty two different pathotypes and pathodemes which can result from five vertical genes. It is important in two ways. If all the known vertical pathotypes and pathodemes of one disease can be fitted into such a model, a gene-for-gene relationship will be established without the necessity for genetical studies in both the host and the pathogen. There may be some gaps due to currently unknown vertical genomes, but their future appearance can then be prophesied with absolute confidence.

Secondly, it has been argued that all vertical pathosystems show a differential interaction but that not only vertical pathosystems do so. It has also been argued that all vertical pathosystems involve a gene-for-gene relationship, and that all gene-for-gene relationships result in a vertical pathosystem. If both these postulations are correct, it would follow that all vertical pathosystems show a Person differential interaction and *vice versa.* Equally, a differential interaction which is demonstrably not a Person differential interaction cannot be due to a vertical pathosystem.

We must now consider Figure 7. If there is a gene-for-gene relationship, the pathotypes and pathodemes can be listed in the same sequence (P1–P32; H1–H32), according to their vertical genomes. The matching interactions then form a symmetrical pattern which can be divided into mirror-images each side of the diagonal from bottom left to top right, and which is the basis of the mathematical properties of this differential interaction. However, these properties are beyond the scope of the present book, as they are somewhat technical and are of interest mainly to specialists who should consult the original. As we shall see in a moment, there is also a short cut which enables us to take them for granted.

Finally, it must be commented that the Person differential interaction was used by Bettencourt and Noronha-Wagner (1967, 1971) to demonstrate the gene-for-gene relationship in coffee leaf rust. As we have seen, genetical studies are impossible in this pathogen as its sexual phase is still unknown.

3.2.7 *The Person/Habgood Differential Interaction*

In the Person differential interaction (Fig. 7) there are five vertical genes and 2^5 genomes (1–32). In the Habgood nomenclature for coffee leaf rust (Fig. 6), there are also five vertical genes and 2^5 genomes (0–31). Person (1959) pointed out that the symmetry each side of the diagonal from bottom left to top right in Figure 7 is due to the fact that the genomes of the vertical pathodemes and pathotypes are listed in the same order. It is a symmetry which is only possible if there is a gene-

Habgood	0	1	2	3	4	0	1	2	3	4	5	6	7	8	9	10	11	12	13	14	15	16	17	18	19	20	21	22	23	24	25	26	27	28	29	30	31
GENES 2^n 0							2		2		2		2		2		2		2		2		2		2		2		2		2		2		2		2
GENES 2^n 1								2	2			2	2			2	2			2	2			2	2			2	2			2	2			2	2
GENES 2^n 2										2	2	2	2					2	2	2	2					2	2	2	2					2	2	2	2
GENES 2^n 3														2	2	2	2	2	2	2	2									2	2	2	2	2	2	2	2
GENES 2^n 4																						2	2	2	2	2	2	2	2	2	2	2	2	2	2	2	2
0						■																															
1	2					■	■																														
2		2				■		■																													
3	2	2				■	■	■	■																												
4			2			■				■																											
5	2		2			■	■			■	■																										
6		2	2			■		■		■		■																									
7	2	2	2			■	■	■	■	■	■	■	■																								
8				2		■								■																							
9	2			2		■	■							■	■																						
10		2		2		■		■						■		■																					
11	2	2		2		■	■	■	■					■	■	■	■																				
12			2	2		■				■				■				■																			
13	2		2	2		■	■			■	■			■	■			■	■																		
14		2	2	2		■		■		■		■		■		■		■		■																	
15	2	2	2	2		■	■	■	■	■	■	■	■	■	■	■	■	■	■	■	■																
16					2	■																■															
17	2				2	■	■															■	■														
18		2			2	■		■														■		■													
19	2	2			2	■	■	■	■													■	■	■	■												
20			2		2	■				■												■				■											
21	2		2		2	■	■			■	■											■	■			■	■										
22		2	2		2	■		■		■		■										■		■		■		■									
23	2	2	2		2	■	■	■	■	■	■	■	■									■	■	■	■	■	■	■	■								
24				2	2	■								■								■								■							
25	2			2	2	■	■							■	■							■	■							■	■						
26		2		2	2	■		■						■		■						■		■						■		■					
27	2	2		2	2	■	■	■	■					■	■	■	■					■	■	■	■					■	■	■	■				
28			2	2	2	■				■				■				■				■				■				■				■			
29	2		2	2	2	■	■			■	■			■	■			■	■			■	■			■	■			■	■			■	■		
30		2	2	2	2	■		■		■		■		■		■		■		■		■		■		■		■		■		■		■		■	
31	2	2	2	2	2	■	■	■	■	■	■	■	■	■	■	■	■	■	■	■	■	■	■	■	■	■	■	■	■	■	■	■	■	■	■	■	■

Fig. 8. The Person/Habgood differential interaction

for-gene relationship. Any order of genomes may be chosen provided that they are both the same. In Figure 8, the arrangement of the genomes is according to the Habgood nomenclature and we can call this the Person/Habgood differential interaction.

It will be observed that it has a greatly increased symmetry, expressed in five basic properties which are exhibited when one vertical gene is added to the system, regardless of how many vertical genes may already be present. With the addition of one gene, the number of vertical genomes is doubled; the number of matching interactions is trebled; the number of both matching and non-matching interactions is quadrupled; the proportion of matching to non-matching interactions is reduced by a factor of 0.75; and the basic pattern is repeated, with a mirror-image symmetry each side of the diagnonal from bottom left to top right.

It was mentioned earlier that the mathematical properties of the Person differential interaction (Fig. 7) were of interest mainly to specialists. With a Person/Habgood differential interaction, these properties may be immediately assumed. It will be seen also that the two genomes which possess no vertical genes are essential components of this symmetrical system and, on these grounds alone, it is legitimate to refer to these genomes as vertical pathotypes and pathodemes respectively.

3.2.8 The Probability of Successful Allo-Infection

We have seen that for every vertical gene added to the Person/Habgood differential interaction, the proportion of matching to non-matching interactions is reduced by a factor of 0.75. With two genes, the proportion is 0.56; with three genes, it is 0.42; with four genes, it is 0.32, etc. Figure 9 shows the proportions of

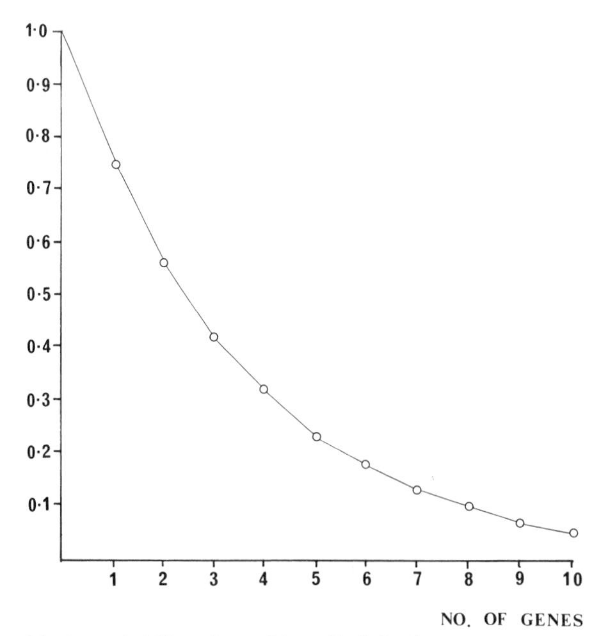

Fig. 9. Probability of matching allo-infection

matching interactions which occur with up to ten vertical genes. It will be observed that this graph is a falling exponential curve and that, with more than eight genes, the proportion of matching interactions is reduced to less than one tenth of the total interactions.

If the ideal vertical pathosystem were an exact replica of the Person/Habgood differential interaction, these proportions would become mathematical probabilities. That is, in any pathosystem with more than eight vertical genes, the probability of successful allo-infection, allowing for all overlapping vertical pathotypes, would be less than 0.1. In reality, of course, no vertical pathosystem is likely to be such an exact replica, but this theoretical treatment does indicate an order of magnitude. From our general knowledge of vertical resistance, we can assume that every vertical pathosystem possesses at least eight vertical genes. The natural pathosystem is balanced and its gene distribution is likely to approximate the Person/Habgood differential interaction. Consequently, it is probably safe to conclude that the overall effects of a natural, vertical pathosystem are at least a tenfold reduction in matching allo-infection.

3.2.9 Agricultural Significance

The discovery of the gene-for-gene relationship was correctly hailed as a major achievement. It was a discovery of great simplicity and great beauty, explaining much that was previously confused. However we must recognise that, so far, this discovery has contributed little to the practical prevention of crop loss. Indeed, it may have contributed to an increase in crop loss because its beauty attracted so much research which might have been directed to more practical ends. It also greatly strengthened an already powerful trend to concentrate on vertical resistance, a trend which became a fashion, increasing until, like most fashions, it reached absurdity.

3.3 Vertical Pathosystem Behaviour

A clear understanding of the functions of vertical resistance in wild plant populations is essential if it is to be successfully ultilised and exploited in agriculture. It must be remembered that vertical resistance can only reduce the exodemic. However, the exodemic is often a major component of the epidemic, and vertical resistance is then very valuable. At this point, it should be taken as axiomatic that, in the natural pathosystem, vertical resistance only occurs heterogeneously; that is, there is always a mixture of different vertical resistances within a wild plant population. In discussion, it is consequently important to make a clear distinction between the population and the individuals within that population. It is also necessary to emphasise that the following discussion is concerned primarily with the effects of vertical resistance, rather than with the resistance itself.

3.3.1 The Breakdown and Recovery of Vertical Resistance

A breakdown of vertical resistance means that the effects of that resistance are lost. In a wild population, a breakdown occurs in an individual rather than in the population as a whole. The plant esodemic begins when the host individual is infected by a vertical pathotype which matches the vertical resistance of that host individual. The epidemiological effect of vertical resistance is thus to reduce the initial inoculum. Of all the pathogen spores deposited on a host individual, only a few are able to infect it, but if the matching vertical pathotype is rare in the pathogen population, a host individual may escape entirely. Once successful allo-infection does occur, however, the esodemic begins and the vertical resistance is valueless against auto-infection. It is still valuable against further allo-infection, and multiple allo-infection is still reduced, but this effect is usually small in the face of massive auto-infection. In a wild host population, the rarity of any vertical pathotype is clearly related to the degree of heterogeneity of that population with respect of different vertical resistances.

A recovery of vertical resistance is the converse of a breakdown, occurring when the consequences of a matching allo-infection, that is, the esodemic, cease completely. This normally happens when there is no further pathogen reproduction or auto-infection because there is no host tissue left. An obvious example is that of a leaf disease in a deciduous tree, where the esodemic ceases with leaf-fall and, in the following season, the new leaves can only be allo-infected. The one tree still possesses the same vertical resistance which has now recovered in the sense that it is still as effective against allo-infection as it was before its breakdown in the previous season. The term recovery is used in the same sense as breakdown to describe the effects of vertical resistance rather than the resistance itself, which clearly remains unaltered while the parasite population changes.

A useful rule is that when vertical resistance breaks down the esodemic begins; when the esodemic ends, vertical resistance recovers.

Because the vertical resistance of any host individual is liable to breakdown, it follows that a mixture of vertical resistances can have population survival value only if the number of recoveries is equal to the number of breakdowns. In a wild population, therefore, there is a regular cycle of breakdown and recovery of vertical resistance.

3.3.2 The Need for Discontinuity of Host Tissue

Vertical resistance is effective if the allo-infection does not match. On a population basis, breakdowns of vertical resistance are thus reduced by a discontinuity of vertically resistant host tissue in space (spatial discontinuity). A recovery of vertical resistance depends on the cessation of the esodemic and the consequent necessity of a new allo-infection. On a population basis, recoveries of vertical resistance are thus dependent on a discontinuity of vertically resistant host tissue in time (sequential discontinuity).

3.3.3 Spatial and Sequential Discontinuity

Spatial discontinuity of vertically resistant host tissue is essentially the same as a plant pattern of vertical resistances (4.4). Within a given area of host population, every individual has a different vertical resistance, which means making allowances for overlapping vertical genomes, that an individual host is not normally infected by its neighbours, nor does it infect them. Epidemiologically, the effect of spatial discontinuity is that of a reduction in population density of both the host and the pathogen. It is the equivalent of the mixed crops of peasant farms, each vertical pathodeme representing a different host species and each vertical pathotype a different pathogen species. Individual breakdowns of vertical resistance are less frequent as a result.

Sequential discontinuity of vertically resistant host tissue means that there are regular periods when there is no host tissue available for infection, and the pathogen is compelled to find an alternative method of survival which is usually some mechanism of over-wintering or aestivation. When host tissue is again available, there is a recovery of its previously broken down vertical resistance, and the pathogen is compelled to produce a wide variety of vertical pathotypes if it is to overcome any one vertical resistance. It normally does this during its sexual phase which permits a wide recombination of vertical genes.

The pathogen usually survives sequential discontinuity of host tissue in the form of resting spores, as in the powdery and downy mildews, smuts, etc.; or by parasitising an alternate host, usually an evergreen perennial, as in many heteroecious rusts; or it may survive as a facultative saprophyte.

If a host population is to obtain the full survival value of vertical resistance, it must exhibit combined spatial and sequential discontinuity of vertically resistant host tissue. Consider a seed-propagated, annual, host population. Spatial discontinuity is maintained by cross-fertilisation and segregation; the population is heterogeneous. If the pathogen is wind-dispersed, as in the rusts, the frequency of allo-infection will be reduced. If the disease is seed-borne, as in the covered smuts, sequential discontinuity is maintained by the fact that each seed is likely to possess a vertical resistance different from either of its parents. If the disease is soil-borne, as in the Fusarium wilts, an individual host growing in a particular plant site is likely to possess a vertical resistance different from its predecessor. When the host population grows in the subsequent season, there is a high probability of every individual possessing an effective vertical resistance to all these pathogens.

3.3.4 Continuity of Host Tissue

We must now consider the converse situation in which a host population has continuity of host tissue in both space and time, for example in the giant grasses, such as sugarcane. At every node there is a leaf and an axillary bud. With age, the leaves are shed and the stem falls over. Any axillary bud which is close to the soil sprouts both leaves and roots and a new plant develops vegetatively. A wild population of such a grass may thus consist of a single, evergreen clone with complete continuity of host tissue in both space and time. In such a host population vertical resistance cannot have survival value; the esodemic is continuous; horizontal resistance is essential and vertical resistance valueless because although it can break down it cannot recover. It is no accident, therefore, that all known resistance to sugarcane diseases is horizontal (6.6).

3.3.5 The Nature of the Host Population

There is clearly a spectrum between the extremes of minimum and maximum continuity of host tissue. We can assess the amount of continuity of host tissue in a wild or progenitor host population and, hence, the probability of its possessing vertical resistance.

A severe winter is more likely to produce a complete cessation of the esodemic than a dry season in the tropics. In the wet tropics, growing conditions are continuous. Vertical resistance should accordingly be more common in temperate than in tropical hosts, and more common in dry tropical than in wet tropical hosts.

Similarly, there is a complete cessation of the esodemic in annual hosts and in the leaves of deciduous trees. Vertical resistance should accordingly be more common in annuals than in perennials, and in deciduous than in evergreen perennials. In perennials, it should also be more common against parasites of leaves, which can be shed, than against parasites of roots, tubers and stems, and pathogens which cause systemic diseases.

Similarly also, the host reproductive method governs continuity of host tissue. Most perennials are out-breeders and this results in heterogeneous populations. Peaches and arabica coffee are obvious exception which, however, probably have sufficient cross-pollination to ensure heterogeneity in a wild population. Most inbreeders are annuals but, again, they have sufficient cross-pollination to provide heterogeneity in a natural pathosystem. An obligate inbreeder or an obligately apomictic reproduction is an evolutionary dead end, equivalent to an exclusively vegetative reproduction. Such reproduction is rare.

We conclude, therefore, that vertical resistance is least likely to occur in wet tropical, evergreen perennials which possess a natural vegetative reproductive mechanism. These include arrowroot, banana, cassava, ginger, pineapple, sisal, sugarcane, sweet potato and yams. Vertical resistance is unlikely to occur in other tropical perennials such as avocado, betel nut, cashew, clove, cocoa, coconut, coffee, date, guava, kapok, mango, nutmeg, oil palm, papaw, quinine, rubber and tea. It is likely to occur in tropical annuals such as maize, rice, sorghum and millets, and most likely to occur in temperate annuals.

Finally, we must note some apparent exceptions. These include leaf rust of coffee *(H. vastatrix)*, South American leaf blight of rubber *(Microcyclus ulei* syn. *Dothidella ulei)* and Solanum potatoes. As we have seen, coffee leaf rust has a pathogen-induced leaf fall during the dry season. This is probably true also of South American leaf blight of Hevea rubber which is less destructive in climates with a marked dry season. There may also be polyphyletic pathosystem (see Chap. 7) but the available data are inadequate to resolve this point. The Solanum potato is a tropical perennial which may, however, behave as a functional annual with respect to its aerial parts. Vertical resistance also occurs to some of its root and tuber diseases such as nematodes and *Synchytrium endobioticum*. This is discussed below.

3.3.6 The Nature of the Parasite Population

In a natural pathosystem, some parasites are more able than others to overcome discontinuity of host tissue. This is a phenomenon which has not attracted much study and further investigation might prove rewarding.

It is safe to assume that vertical resistance is unlikely to evolve when its survival value is low, and we can conclude that vertical resistance should be more common against some categories of parasite than against others. Let us begin by comparing insects and fungi. We note that vertical resistance to insects is rare, a gene-for-gene relationship having been demonstrated only once, with the Hessian fly in wheat, while vertical resistance to fungal parasites is common.

An essential feature of plant parasitism is that the parasite and the host must come into contact, which can happen in three ways. Either the parasite finds the host; or the host finds the parasite; or the contact is never broken, as with some seed-borne diseases. If the parasite has to find the host, it must possess a long-distance dispersal mechanism, such as the wind-borne spores of fungal parasites and with most insect parasites. If the parasite has no such dispersal mechanism, it must wait, usually quite dormant, until the host happens to come into contact. This occurs typically with many soil-borne fungal parasites and with root nematodes. It is a feature of such parasites that they often possess dormancy mechanisms of great longevity. The resting spores of potato wart disease (*S. endobioticum*) and the cysts of *Heterodera* spp. can remain dormant but viable for several decades. It is also likely that such mechanisms are commonest in parasites of annual hosts in which the host population changes with every season.

There are several key features of the long-distance dispersal of fungal parasites. Firstly, the dispersal is random, so that the wastage of spores is enormous and they must be produced in vast numbers. In its turn, this means both that the spores must be very small and that the majority of them must be produced asexually. Characteristically, a fungal parasite undergoes sexual recombination at the end of the epidemic, prior to over-wintering or aestivation. The sexual phase produces large numbers of spores of many vertical pathotypes, thus increasing the exodemic. Once the esodemic begins, asexual reproduction occurs, thus increasing auto-infection. Some parasexual changes and mutations can occur during the esodemic but these are relatively rare when compared with the numbers of asexual spores. This pattern occurs with some insects and it is tempting to suggest that the function of sexual reproduction in aphids is to increase the exodemic, while that of parthenogenetic reproduction is to increase the esodemic.

Insects differ from fungal spores in that their long-distance dispersal is non-random. They possess tactical mobility and can respond to external stimuli. This response is feedback and insects possess the properties of a simple machine designed to illustrate the fundamentals of cybernetics. (It is notable that insect parasites operate entirely with inherited behaviour patterns. They usually cannot learn and they lack intelligence, that is, the ability to solve new problems. However, their life-span is too short for acquired behaviour patterns to have survival value and new problems are very uncommond in a natural pathosystem). The main effect of tactical mobility is that an individual parasite can both search for and find its host, and because the parasite dispersal is non-random, the wastage of individuals is minimal. In its turn, this means that the individuals need to be produced in relatively small numbers; they can be considerably larger than fungal spores; and sexual recombination can occur with greater frequency. In general, therefore, insects are more able to overcome discontinuity of host tissue than are the fungi.

It is just possible that some insect parasites could develop the ability to identify and avoid a non-matching vertical resistance. As we have seen, vertical resistance is rare against insect parasites and common against fungal parasites.

Vertical resistance against bacterial parasites is rare. It apparently occurs in *Phaseolus vulgaris* against halo blight *(Pseudomonas medicaganis phaseolicola)* but this seems to be the one exception which proves the generality of the rule. The reason for this rarity is not clear but it may be associated with the very short reproductive cycle which permits vast numbers of individuals and, hence, large numbers of variants, to be produced very rapidly.

Vertical resistance against viruses does not seem to occur. Some differential interactions have been demonstrated but it seems that no Person differential interactions are known. Again, the reason is not clear but it may be that the capacity of viruses to make the simple change necessary to overcome vertical resistance is too great for this kind of resistance to have survival value against them.

3.3.7 The Crop Pathosystem

So far, the discussion has concerned the natural pathosystem. A population of parasites is dispersed through an ecosystem in which the host population is mixed. Many useless parasite contacts are made. The useless contact may involve a non-living surface, or a non-host, or a non-matching, vertically resistant host. The host population is non-uniform and there is discontinuity of host tissue.

The crop pathosystem is very different. The essential feature of agriculture is high productivity; that is, the maximum production per unit of human labour. This is typical in the wheat lands of North America: huge fields, gangs of combine harvesters, and bulk storage, transport, marketing and processing all demand uniformity of product and, hence, uniformity of crop. There is a tendency among prophets of doom to deplore crop uniformity, which is quite wrong. Crop uniformity is essential and, if we can only manage our crop pathosystems by abandoning crop uniformity, we must confess to abject and humiliating defeat.

It is common knowledge that the value of vertical resistance in crop pathosystems varies enormously, from the valuable to the totally valueless. We must now enquire what factors govern the value of vertical resistance in agriculture. In the next Chapter, we shall examine possible methods of enhancing this value.

Robinson (1971), drew up a set of rules designed to assist in assessing the usefulness of vertical resistance to any given pathogen. The common feature of these rules, the relative flexibilities of the host and pathogen populations, only become apparent subsequently. The flexibility of the host population dynamics determines the combined spatial and sequential discontinuity of host tissue under agricultural conditions. The flexibility of the parasite population dynamics determines the ability of the parasite to overcome that discontinuity. It will become apparent that the value of vertical resistance is high when the host flexibility is maximal and the pathogen flexibility is minimal. If the host flexibility is reduced and/or the pathogen flexibility is increased, vertical resistance becomes increasingly less valuable.

3.3.8 *The Phenomenon of Rarity*

If an epidemic is to develop, individuals of the parasite population must come into contact with individuals of the host population. The intensity of the epidemic is dependent on the frequency of those contacts. With a given host population, the frequency of contact is related to the size (i.e. total number of individuals) of the parasite population. If the parasite is rare, contact is also rare and the epidemic is slight. Equally, for a given parasite population, the frequency of contact is related to the size of the host population, and again if the host is rare, contact is also rare and the epidemic is slight.

It is now clear that vertical resistance reduces the epidemic by increasing the rarity of contact, or more correctly, the rarity of successful (matching) contact. In practice, this means that successful management of the vertical pathosystem must depend on two criteria. The rarity of contact must be increased and this is achieved by an increased rarity of both vertical pathodemes and vertical pathotypes. Rarity of vertical pathodemes is achieved by the use of patterns of host populations; and rarity of vertical pathotypes is achieved by the use of strong vertical genes (4.4).

3.4 The Components of Host Population Flexibility

The following list of components of host flexibility is not intended to be comprehensive. Seven of the more important components are described but many others doubtless occur, particularly those of isolated but special importance in some of the less known pathosystems.

3.4.1 *Ease of Cultivar Replacement*

When the vertical resistance of a cultivar breaks down, it is normally replaced with a new one with a different vertical resistance. This is the essence of the boom-and-bust cycle of cultivar replacement which should be regarded as both an uncontrolled and a somewhat clumsy method of achieving sequential discontinuity of host tissue. Cultivar replacement is obviously easier with annual than with perennial crops; therefore we conclude that vertical resistance will be more useful in annual crops.

There are two prerequisites of cultivar replacement; a new cultivar must be available and in an adequate quantity. This means that there must be a breeding team engaged in repetitive breeding (9.3) and an efficient seeds industry which can both multiply and distribute the cultivar effectively. Both these factors mean that vertical resistance is likely to be less valuable in the less developed countries where such agricultural infrastructure is either deficient or lacking.

3.4.2 *Ease of Breeding*

Some crops are easier to breed than others and consist of thousands of cultivars whereas others consist of relatively few cultivars. Ease of breeding is related to many factors such as the overall genetic variability, the reproductive method, etc. It is also related to the crop quality requirements. Quality in wheat, for example, is relatively unimportant compared with the highly valued and specialised qualities of some horticultural cultivars. In some crops, such as tomatoes, breeding is so easy that there is often a cultivar boom-and-bust cycle irrespective of the breakdown of vertical resistance.

The ease of breeding and, hence, the number of available cultivars is thus a component of host flexibility.

3.4.3 Availability of Vertical Genes

Cultivars can only be easy to replace and breed, with respect to their vertical resistance, if an adequate number of vertical resistance genes is available. It is most unlikely that any host species has an inexhaustible supply of vertical genes, although it may prove possible to produce genuinely new ones by induced mutation. More important, however, is the availability of strong vertical genes, which are, as a rule, rare. This important topic is discussed in the next Chapter. The value of vertical resistance is higher in crop species which possess adequate numbers of strong vertical genes.

3.4.4 Population Size and Uniformity

Relatively small crop populations involving many different vertical pathodemes clearly provide a higher host flexibility than one large population of one vertical pathodeme. We must conclude that vertical resistance will prove more valuable in small and varied crop populations.

3.4.5 Cultivar Popularity

Van der Plank (1968) has pointed out that the value of a vertical pathodeme is inversely proportional to its popularity as a cultivar. Increasing popularity constitutes the boom phase of a cultivar boom-and-bust cycle. The greater the boom, the more damaging and, indeed, the more probable is the bust. Cultivar popularity is thus a reduction in host flexibility, while its converse is a balanced cultivar competition. Popularity implies that all other cultivars are inferior; they lack competitiveness while the popular cultivar has an excessive, unbalanced competitiveness. Host flexibility is reduced by an excessive cultivar popularity.

3.4.6 Ease of Exploitation of Patterns of Vertical Resistance

We have seen that, in nature, the sole function of vertical resistance is to reduce the frequency of matching allo-infection in a genetically heterogeneous host population. The demands of commerce and economics insist that agricultural products are uniform. In some crop species it is possible to exploit patterns of different vertical resistances without sacrificing uniformity of product. The nature of these patterns is discussed in the next Chapter. It is clear, however, that a move towards the mixed vertical resistances of nature will enhance the value of vertical resistance in agriculture. If such a move is easy, the crop in question has a higher flexibility than when it is difficult.

3.4.7 *Complex Vertical Pathodemes*

There is some evidence that vertical and horizontal pathogenicities are inversely correlated. That is, a complex vertical pathotype will have a lower horizontal pathogenicity than a simple vertical pathotype, from which it would follow that a complex vertical pathodeme can only be attacked by a pathotype with a reduced horizontal pathogenicity. The effect of a reduced horizontal pathogenicity is the same as that of an increased horizontal resistance; there is less disease. The esodemic of a complex vertical pathodeme would be less damaging than that of a simple vertical pathodeme. The theoretical possibility has still to be demonstrated, but if it is established it may lead to new methods of exploiting vertical resistance. It should be noted that this is another form of apparent horizontal resistance.

3.5 The Components of Parasite Population Flexibility

3.5.1 *Vertical Mutability*

The vertical mutability of a parasite can be defined as the readiness with which it produces new vertical pathotypes, regardless of whether such production is by mutation, sexual or parasexual means or by the population increase of a previously rare pathotype. It is not possible to measure vertical mutability accurately, but a useful, comparative scale can be drawn up on the basis of the size of population of a given vertical pathodeme and the time taken before its vertical resistance breaks down. We assume that a given number of parasite individuals must be produced for a particular vertical pathotype to appear. That number of individuals is a function of both time and the overall size of the parasite population.

P. polysora of maize clearly has a very high vertical mutability. Storey and Ryland (1955) found that vertical pathodemes of maize intended as new cultivars in Kenya, broke down to new vertical pathotypes within one season and that this occurred in both breeders' plots and the very small maize populations within a research glasshouse. Vertical resistance to this pathogen has never been successfully employed in agriculture. *P. infestans* has a considerably lower vertical mutability. The potato cultivar "Pentland Dell", which possessed R-genes 1, 2 and 3, remained free from blight for three years of commercial production in Britain before its vertical resistance broke down. This level of vertical mutability is still too high to justify the use of vertical resistance in agriculture. "Pentland Dell", after all, required some ten years of testing and multiplication prior to its commercial release to farmers. The resistance of a cultivar which lasts only a fraction of the time taken to produce it has little agricultural value. It should be added that, in Mexico, where *P. infestans* forms functional oospores, its vertical mutability is so high that vertical resistance is of no agricultural value whatever.

The vertical mutability of *Puccinia graminis tritici* is relatively low. A vertical pathodeme of wheat can often be grown in millions of acres for 10–20 years before a matching vertical pathotype appears. Possibly the lowest vertical mutabilities are those of *F. oxysporum* f. sp. *conglutinans* and *S. endobioticum*. Vertical resistance to these two pathogens has proved very valuable.

3.5.2 Reproductive Capacity

A pathogen with a high reproductive capacity can obviously produce a population explosion of a new vertical pathotype far more quickly than one with a low reproductive capacity. Possibly the highest fungal reproductive capacity known is that of *U. scitaminea*, the cause of sugercane smut. Waller (1969) has estimated that one smut whip can produce 10^{11} spores.

3.5.3 Dissemination Efficiency

Consider potato blight *(P. infestans)* and potato wart disease *(S. endobioticum)*. If a new vertical pathotype of blight appears, it can spread for hundreds of miles in the course of one season. Blight spores are wind-borne and have a high dissemination efficiency. Wart disease, on the other hand, has a low dissemination efficiency, and the lower the dissemination efficiency, the lower the pathogen population flexibility. Vertical resistance is clearly more valuable when a new vertical pathotype spreads so slowly that there is time to identify it and, possibly, to contain or even eradicate it with isolation and quarantine measures.

3.5.4 Legislative Control

In Western European countries, the cultivation of potatoes is legislatively controlled in relation to wart disease. Any field in which wart disease occurs is scheduled, and it is then illegal to use that land for any potato seed production or for the cultivation of any wart-susceptible potato cultivar. The resting spores can then germinate but only in the presence of a vertical pathodeme which they do not match. *S. endobioticum* has a low vertical mutability and, under these circumstances, is unable to produce new vertical pathotypes. This legislative control has so reduced the pathogen population flexibility that vertical resistance provides a complete and lasting control of wart disease. Indeed, wart disease will eventually be totally eliminated from these countries, even though the resting spores remain viable for at least thirty years and probably much longer. In other parts of the world, notably Newfoundland and parts of Eastern Europe, there is no legislative control, and the pathogen population is correspondingly large and its flexibility is increased. New vertical pathotypes have appeared in these areas and vertical resistance does not provide an adequate disease control.

Wart disease has a low dissemination efficiency and legislative control of its vertical resistance is effective. Such control has not yet been attempted with diseases of high dissemination efficiency such as the wheat rusts and potato blight, a possibility discussed in the next Chapter.

3.5.5 Degree of Protection

Consider potato blight *(P. infestans)* and wheat stem rust *(P. graminis)*. Vertical resistance to blight confers complete protection against non-matching vertical pathotypes. This means that a potato population carries no blight whatever for as long as its vertical resistance endures. A new vertical pathotype can only arise in another blight population parasitising another potato population. Vertical resistance to stem rust, however, does not necessarily confer complete protection against non-matching vertical pathotypes; it is often incomplete vertical resistance. This means that a vertical pathodeme may carry some rust, which increases the flexibility of the pathogen population. A rust population, however limited it may be, exists on the vertical pathodeme and a matching vertical pathotype may arise within it, and the vertical resistance is liable to breakdown more quickly than if complete protection were conferred. It must be remembered that a single vertical pathodeme of wheat may be grown in millions of acres. Although the incomplete vertical resistance may provide an agriculturally adequate control of rust, the rust population in those millions of acres is nevertheless of considerable size. Furthermore, in the event of a matching vertical pathotype being produced in this rust population, the pathotype is ideally located for a rapid population explosion in the matching wheat pathodeme.

3.5.6 Disease Transmission

Some diseases are transmitted by the propagating material of the host. Potato seed tubers, for example, can carry blight *(P. infestans)*, and given suitable weather conditions, one blighted seed tuber in about 100,000 tubers provides an adequate initial inoculum for a major blight epidemic. If the potato cultivar in question has vertical resistance to blight, the blighted tubers will be carrying a matching vertical pathotype. The flexibility of the pathogen is then increased accordingly and the vertical resistance has broken down before the epidemic has even started. The esodemic is continuous and vertical resistance cannot recover because there is no sequential discontinuity of host tissue.

3.5.7 Closed Seasons

If there is a closed season, such as a severe winter in the temperate regions, or a severe, long, dry season in the tropics, pathogen populations decline to near-extinction. More important, the population of a particular vertical pathotype may become extinct, a phenomenon which can be deliberately exploited with soil-borne diseases by crop rotations which create artificial closed seasons. When there is no closed season, however, the flexibility of the pathogen population is greatly enhanced. In Kenya, for example, there is no winter and commercial wheat crops exist for ten months of each year. The remaining two months are a dry season which is too short to destroy rogue wheat plants carrying rust and, more particularly, a given vertical pathotype. The epidemic never really dies out and, more specifically, the esodemic is maintained until a cultivar is discarded. The average commercial life of a vertically resistant wheat cultivar in Kenya is only 4.4 years, which is less than the time required to breed it.

3.5.8 Dormancy Mechanisms

If a parasite possesses a long-term dormancy mechanism such as the resting spores of *S. endobioticum* or the cysts of *Heterodera rostochiensis*, a vertical pathotype can be preserved for several decades, thus increasing the pathogen flexibility and precluding any possibility of a rotation of different vertical pathodemes in which old pathodemes are re-used after a reasonable interval.

This kind of dormancy mechanism enables the pathogen to overcome a natural sequential discontinuity in host tissue which is exceptionally long-termed. It occurs when the pathogen population is essentially non-mobile and it is the host individual which encounters the parasite, in the course of host population changes during periods of many years.

Chapter 4 Vertical Pathosystem Management

4.1 Conventional Management

4.1.1 Historical Outline

The scientific breeding of plants began some seventy years ago. The breeders, understandably, were primarily interested in the positive aspects of breeding: increased yield and increased quality. They have some outstanding successes on record, including the development of hybrid maize, soya breeding and, more recently, the miracle wheats and rices of the green revolution. Breeding for resistance to pests and diseases was always considered somewhat negative, involving the prevention of crop loss, rather than the promotion of crop gain.

As we have seen, there was an intellectual boundary which separated the discipline of plant breeding from the parasitological disciplines of plant pathology, entomology and nematology. The prevention of crop loss was regarded primarily as the parasitologists' responsibility, and after all, the pathologists, in particular, had some outstanding (and much earlier) successes on record. The use of sulphur against powdery mildews and Bordeaux mixture against downy mildews for example produced results quite as dramatic as the use of hybrid maize seed. But there were still some insoluble problems; crop spraying was uneconomic on extensive crops such as the cereals and flax; and neither copper nor sulphur would control some diseases such as wart disease of potato. Clearly, breeding for resistance was necessary.

It was also exciting; Mendel's laws had been rediscovered, the role of chromosomes demonstrated, and gradually, a set of breeding principles was developed. The first step was to find a "source" of resistance which, obviously must confer complete protection. Preferably, this resistance should be due to a single mechanism which could be easily recognised and easily studied, but even more important, it should be inherited by a single gene which could be easily manipulated genetically. It was the breeders' task to transfer that gene to a cultivar with good yield and quality. It was then up to farmers to grow the resistant cultivar. The whole thing was really very simple; given enough plant breeders, there would be no further need for pathologists or entomologists.

Initially, this approach was apparently justified, and we can consider only the potato crop, remembering that similar stories can be told of other crops. Wart disease *(S. endobioticum)* was threatening the British potato crop with ruin, but the cultivar "Snowdrop" was apparently immune. This was due to its hypersensitivity mechanism, a phenomenon which attracted a great deal of attention, and its inheritance was controlled by a single gene. It was true that there was some "physiologic specialisation" but there were also several different resistance genes which, together, covered the spectrum of this specialisation. New cultivars were produced, and the legislators, with the food shortages of World War I fresh in their minds, acted promptly. There was a register of varieties and variety synonyms were abolished. Seed certification was introduced. Contaminated fields were scheduled, and their use for either seed-production or susceptible varieties was prohibited; the problem was solved. It is true that the resting spores remain viable for a very long time, but there is every reason to believe that wart disease will eventually become extinct in Britain.

It is difficult now to assess the influence that this success had on potato breeders. The same approach was tried with other diseases, including blight *(P. infestans)*. The problem was less urgent because blight could be controlled by spraying, but it would be better to avoid the cost of spraying. The source of resistance was the wild *Solanum demissum* which was apparently immune. It too had a hypersensitivity mechanism whose inheritance was controlled by a single gene. There was great optimism but it was premature, and blight is still controlled by spraying.

Simmonds (1969) has discussed the history of potato breeding. He comments that the 19th century, amateur potato breeders were immensely successful, many of the varieties produced more than sixty years ago still being of agricultural importance. About 1910, however, scientific research began to have an increasing influence, and the private potato breeder was replaced by the large-scale, professional team. During the late 1960's, it was estimated that about six million potato seedlings were being screened annually in northern temperate countries. We can only guess at the total number of seedlings bred and screened during the past sixty years; it may be 50–100 million. And yet, old, pre-1910 varieties, produced by amateurs are still agriculturally important. Simmonds (1969) quotes Russel-Burbank, Bintje Majestic and King Edward as examples. There is no question that potato breeding during the last 60–70 years has produced results; on that scale, it could hardly fail to do so, but given the enormous research resources devoted to it, it has been conspicuously unsuccessful. Simmonds (1969) suggests various reasons for this comparative failure, one of the main ones being the wasted years devoted to vertical resistance to blight. He comments also that, by 1969, no single disease-resistance gene (i.e. vertical resistance) derived from wild or primitive potatoes had produced a significant impact on potato production and he questions the contribution made in this connection by the potato collections of the world.

4.1.2 The Influence of the Breeding Technique

A review of the literature on the inheritance of disease resistance in plants shows an almost universal preoccupation with oligogenically-inherited resistance and Mendelism. Biffen (1905) started the trend with his classic paper entitled "Mendel's laws of inheritance and wheat breeding". Nearly 70 years later, Person and Sidhu (1971) traced 912 papers on the genetics of plant disease resistance and showed that, while 875 papers reported resistance due to major genes, only 60 papers reported resistance due to minor genes. It is now clear (Robinson, 1973a) that some 70 years of plant breeding for disease resistance have been dominated by the convenience of the breeding technique, regardless of the plant pathological consequences of that technique. These consequences were assumed to be normal, natural and inevitable. Pedigree breeding favours oligogenically inherited characters which are easy to manipulate in a breeding programme, easy to measure and study, and convenient to describe for publication; but oligogenically inherited resistance is almost invariably vertical, and therefore temporary resistance.

4.1.3 The Boom-and-Bust Cycle

Temporary resistance leads to a boom-and-bust cycle of cultivar production, and we must now consider its advantages and disadvantages. The main advantage is that, while it lasts, vertical resistance usually provides a complete control of a disease. The critical factor is how long the boom lasts before the bust occurs, and this factor is immensely variable. Vertical resistance to *P. polysora* in Africa broke down so quickly that the bust occurred before the boom had even started. Some 60 years of wheat breeding in Kenya have shown that the average commercial life of a wheat cultivar in that country was 4.4 years. Assuming an average Kenyan cultivation of 0.25 million acres of wheat each year, this means that a new cultivar was produced for roughly every million acres of cultivation. If these figures applied to world wheat production, which, of course, they do not, the world-wide boom-and-bust cycle in wheat would require some 300 new wheat cultivars every year. Perhaps a more useful guide is to compare the life of a vertically resistant cultivar to the time required to breed it.

4.1.4 The Cost of the Boom-and-Bust Cycle

In spite of the enormous amounts of plant pathological research, we still lack adequate techniques to assess crop loss due to disease (Chiarappa, 1971); few accurate assessments are available, and we must accordingly use indirect assessments. The best indirect indication seems to be an index in which the volume of plant pathological publication is related to the value of the crop. However, this approach has three limitations which must be clearly recognised; it produces relative results only, and the volume of publication is not necessarily related to the economic importance of a plant disease. For example, the discovery of the gene-for-gene relationship stimulated considerable research of a somewhat academic nature. A sudden and dramatic disease loss is also more likely to attract research than one which unobtrusively exacts a steady toll. A boom-and-bust cycle, by its very nature, is thus likely to attract more research than a constant disease which causes the same average loss. However, in spite of these limitations, the comparisons are valuable.

Stevens (1939, 1941) first attempted to produce such an index and his demonstration that disease is more important in inbreeders than in outbreeders is well known, using the number of pages of publication per crop per annum, divided by the annual value of the crop. His figures concern the U.S.A. only and are given in Table 1.

More interesting are the data from the review of Person and Sidhu (1972), already discussed. Of the 875 papers reporting oligogenically inherited resistance, 29% concerned wheat and a further 27% concerned barley, oats and flax. In other words, more than half the total publications on the genetics of disease resistance

Table 1. The Stevens Index

Crop	Breeding system	Vol. of publications corrected by value of crop	Estimated loss from disease (%)
Flax	Inbreeder	14.2	
Rice	Inbreeder	4.9	
Barley	Inbreeder	3.5	2.7
Wheat	Inbreeder	3.4	5.2
Sorghum	Inbreeder	2.3	
Oats	Inbreeder	1.8	2.8
Rye	Outbreeder	1.5	
Maize	Outbreeder	0.8	0.4
Buckwheat	Obligate outbreeder	0.0	

Table 2. Inbreeders and outbreeders compared

Comparison	Oats (*Avena sativa*)	Rye (*Secale cereale*)
Pollination	Inbreeder	Outbreeder
Breeding technique	Pedigree	Mass selection
Numbers of plants	Individuals	Populations
Inheritance favoured	Oligogenic	Polygenic
Resistance favoured	Vertical	Horizontal
Disease control	Temporary	Permanent
Total cultivation 1950–1970	832 m/hectares	536 m/hectares
No. of plant pathological papers published/million hectares	3.6	1.2
No. of rust papers published/million hectares	1.2	0.17

concern only four crops, of which three are of only moderate economic importance. It is probably not necessary to comment that these four crops are all seed-propagated, inbreeding annuals which are cultivated as pure lines, each protected by a single vertical resistance. Fifteen other crops, which are seed-propagated, outbreeding annuals collectively accounted for only 14% of the 875 papers. On average, therefore, each of the four inbreeders attracted about fourteen times as much research on disease resistance genetics as each of the outbreeders, wheat alone attracting nearly thirty times as much as the average outbreeder although here there is clearly an indeterminate, economic bias.

Robinson (1973a) attempted a more rigorous comparison using two crops which were as similar as possible in all respects except for their pollination method. Choosing oats and rye, he based his index of the number of plant pathological papers published for each million hectares of world cultivation during the period 1950–1970 (Table 2). The publishing rate was determined from the *Review of Applied Mycology* which can be considered free from bias towards either crop; that is, although this journal does not abstract all published papers the proportion of papers abstracted does not differ between these two crops. The total cultivation of each crop, in millions of hectares, was obtained from FAO Production Yearbooks, 1950–1970. The limits of accuracy of these yearbooks are clearly stated and, once again, the proportion of error is unlikely to differ between oats and rye. It will be seen that there are 3.6 plant pathological papers per million hectares of oats and 1.2 papers per million hectares of rye. We conclude that all plant pathology, including the study of non-parasitic disorders, is three times as important in oats as it is in rye. However, there are 1.2 papers on the rusts of oats compared with 0.17 papers of the rusts of rye, per million hectares. The rusts are primarily responsible for the boom-and-bust cycle of oats and they are apparently seven times as important in oats as they are in rye, in which there is no boom-and-bust cycle.

It would be dangerous to generalise from these figures, but they do indicate an order of magnitude. Plant diseases are several times more important in an inbreeding annual than in an outbreeding annual, the difference being apparently due to the use of vertical resistance and the consequent boom-and-bust cycle of cultivar production.

4.1.5 The Vertifolia Effect

It is normally assumed that, when a vertical resistance breaks down, the cultivar in question must be discarded because it is then far too disease-susceptible to be of any further value. This is in marked contrast to the natural pathosystem in which horizontal resistance provides adequate protection in the esodemic. The cultivar susceptibility is due to a host erosion of horizontal resistance (6.4); that is, horizontal resistance has been lost due to genetical changes in the host population. In a boom-and-bust cycle, there is a special kind of host erosion of horizontal resistance which van der Plank (1963) has named the vertifolia effect, after the potato cultivar Vertifolia which was bred for vertical resistance to blight and which proved exceptionally susceptible when that vertical resistance broke down. The vertifolia effect can thus be defined as a host erosion of horizontal resistance which occurs during breeding for vertical resistance.

It is apparent that the vertifolia effect is an essential feature of a boom-and-bust cycle, because, had there been no host erosion of horizontal resistance, the effects of the breakdown of vertical resistance would be relatively slight and the cultivar would, in all probability, be retained. Indeed, the nature of the boom-and-bust cycle indicates the importance of the vertifolia effect which is so common that it is usually overlooked as a factor of any significance.

4.2. Vertical Pathosystem Demonstration

The demonstration of the vertical nature of a pathosystem is an essential aspect of pathosystem management. There are six categories of evidence.

4.2.1 Mechanisms

Vertical resistance is conferred by mechanisms which are within the capacity for change of the parasite. This is not a characteristic that is easy to demonstrate although a hypersensitivity mechanism is almost certain to confer vertical resistance if it is a normal feature of the pathosystem. At the histological level of the pathosystem, vertical resistance is likely to be conferred by a single, active mechanism which is also relatively simple and one which prevents allo-infection. At the physiological or biochemical level of the pathosystem, however, this single, simple mechanism is likely to involve many complex processes.

4.2.2 The Differential Interaction

This matter has already been discussed. The demonstration of a gene-for-gene relationship by genetic studies in both host and parasite is proof of a vertical pathosystem. So is the demonstration of a Person differential interaction. Other kinds of differential interaction are not due to vertical resistance.

4.2.3 Inheritance

Vertical resistance can only be inherited oligogenically but oligogenic inheritance is not proof of vertical resistance. In this context, it is worth noting that quantitative vertical resistance is always inherited oligogenically and operates in both young and mature plants. Quantitative horizontal resistance is always inherited polygenically and is often at a higher level in mature plants than in seedlings.

4.2.4 Epidemiology

It is so easy for horizontal resistance to be an apparent vertical resistance and *vice versa* that epidemiological evidence is usually too complex to be useful.

4.2.5 The Host

The nature of the host can provide evidence in support of the vertical nature of a pathosystem, but the converse is likely to be far stronger evidence. If, in its natural pathosystem, the host has both spatial and sequential continuity of tissue, vertical resistance has no evolutionary survival value.

4.2.6 The Parasite

As we have seen, vertical resistance is rare against insect, bacterial and virus parasites, but common against fungal and nematode parasites. However, this is no more than supporting evidence.

4.3. Strong Vertical Genes

One of van der Plank's (1968) most penetrating and important insights is his concept of strong vertical genes. Vertical gene strength is important in two ways, determining both the rarity of a vertical pathotype and the rate at which that pathotype becomes rare, after it has been common. It is probably safe to conclude that strong vertical genes are essential for effective, vertical pathosystem management; without them, the management cannot be effective.

4.3.1 Definition

The terms "strong" and "weak" should be regarded as abstract terms without any descriptive function. Just as the term "vertical" can be differently but accurately defined in many contexts, so can the term strong. The only criterion is that the various definitions must make sense and must fit the facts.

In the traditional boom-and-bust cycle, the boom has two important features. It continues only until the vertical resistance breaks down, and while it continues, it exerts positive selection pressure for the matching vertical pathotype. The greater the boom, in both space and time, the greater the selection pressure. The duration of the boom depends on the rarity of the matching vertical pathotype. If it is rare, the boom is a big one; if it is common, the boom may never begin. In the first case, the vertical resistance is described as strong; in the second, it is described as weak.

The bust which follows the boom also has two important features. The bust is due to the fact that the matching vertical pathotype has become common because of the selection pressure in its favour, and it ensures that cultivation of the vertical pathodeme ceases. Selection pressure for the matching vertical pathotype then ceases also, and the pathotype tends to disappear. If it disappears quickly, the vertical resistance is described as strong; if slowly, or not at all, the vertical resistance is described as weak. As we shall see shortly, it is perhaps more accurate to speak of strong and weak vertical genes and genomes.

One point must be emphasized. The concept of strong and weak vertical genes is a *concept*, and a brilliant one. Nevertheless, it has attracted strong opposition. Whatever the concept, is it a fact? There are two considerations. If strong vertical genes cannot be demonstrated in a particular vertical pathosystem, this is negative evidence which, however, does not contradict the concept. If we are to deny the concept, we must show that all the vertical genes of a vertical pathosystem have an equal rarity and an equal rate of becoming rare, and we must demonstrate this for all vertical pathosystems. Conversely, if we wish to prove the concept, we need to demonstrate differences in strength of vertical genes in only one vertical pathosystem. If we wish to prove the generality of the concept, we must do this in many vertical pathosystems, a general demonstration which will probably be made. Secondly, even if the concept, and its generality, are proved, it is still legitimate to dispute its practical value in terms of vertical pathosystem management, a consideration which must now be discussed.

4.3.2 The Mechanisms of Vertical Gene Strength

The best explanation of differences in vertical gene strength is based on the theory that survival values tend to balance; one can be increased only at the expense of another (5.5). It is probable that vertical pathogenicity can be gained only at the expense of one or more other survival values, these reduced survival values having possibly contributed to other aspects of epidemiological competence (8.1).

For example, there is evidence that vertical and horizontal pathogenicity are often inversely correlated. This is easily demonstrated by comparing the relative growth rates of both a simple and complex vertical pathotype on the same simple vertical pathodeme. Let us suppose that this demonstration has been made; the complex vertical pathotype has a reduced horizontal pathogenicity. In competition with a simple pathotype, parasitising a simple pathodeme, the complex pathotype will reproduce more slowly, and will tend to disappear from the pathosystem. We thus return to the systems concept of balance, the vertical gene being unnecessary and representing a loss of balance. One survival value (vertical pathogenicity) is too high and, as a result, another (epidemiological competence) is too low, and competitiveness is reduced. Unnecessary survival values are not values at all, but hindrances and they consequently tend to disappear.

Now let us suppose that *different* vertical genes can be acquired by the parasite only at the expense of *different* survival values. These various reduced survival values may contribute unequally to epidemiological competence. When the reduction is important, the vertical gene is strong; when unimportant the vertical gene is weak.

It follows that we expect two kinds of difference in vertical gene strength. One gene may be stronger than another in one pathosystem and one gene may be stronger in one pathosystem than the same gene in another pathosystem. It is this second point which we must examine.

The gene is strong because it has been gained at the expense of other survival values. The reduced survival values may be more important in one environment than in another, therefore it is likely, that a gene which is strong in one part of the world will be weak in another. Because the environment is variable, it is possible that a gene which is strong in a normal season will be weak in an abnormal season in the same locality. The reduced survival values may contribute to parasitism, being more important in one pathodeme than another. It is likely, therefore, that a vertical gene may be stronger in one cultivar than in another. (For example, a reduced horizontal pathogenicity will be a greater hindrance to a parasite in a cultivar with high horizontal resistance than in one with low horizontal resistance). It follows that we must expect the strength of a vertical gene to vary, between localities, between seasons and between cultivars, and such variation would account for apparent conflicts in experimental data. However, it remains to be seen how important this variation proves to be in practical, vertical pathosystem management.

So far, we have been discussing the strength of individual vertical genes. We must now consider the strength of vertical genomes, which may include more than one strong gene. There are several possible effects of a combination of strong vertical genes in one genome. The amount of variation in strength (between localities and/or seasons and/or cultivars) may be either reduced or increased. Clearly, this would depend on whether the several strong genes were all gained at the expense of the same or of different survival values in the parasite. More important, we would expect the strength of a genome possessing several strong genes to be greater than that of a genome possessing fewer or only one strong gene, the cumulative effects being either additive or synergic (i.e. the total being greater than the sum of the parts).

Most intriguing of all, however, is the possibility that the cumulative effects of several strong genes may lead to a Hegelian change, a small difference in degree producing a difference in kind. We assume that vertical pathogenicity can only be gained at the expense of one or more of the other survival values which may contribute to epidemiological competence. If any one of those values is reduced beyond its critical point, epidemiological competence will be lost entirely. The matching vertical pathotype may appear, possibly repeatedly, but it would not survive. The vertical resistance would then be permanent, at least in that locality, in a normal season and in that particular cultivar. It seems that this exciting possibility has now been demonstrated in wheat stem rust (4.5).

It is these and similar considerations which led van der Plank (1975) to formulate the second gene-for-gene hypothesis. He argues that Flor's (1942) hypothesis is concerned with gene identity. The second hypothesis is concerned with gene quality. It is stated as follows: "In host-parasite systems in which there is a gene-for-gene relationship, the quality of a resistance gene in the host determines the fitness of the matching virulence gene in the parasite to survive when this virulence is unnecessary; and, reciprocally, the fitness of the virulence gene to survive when it is unnecessary determines the quality of the matching resistance gene as judged by the protection it can give to the host". [Van der Plank (1975) uses the word "virulence" to mean vertical pathogenicity.] That is, the greater the value of the gene to the host, the less its value to the parasite, and *vice versa.*

In the most simple, practical terms, this means that we must not only identify vertical genes but we must also measure their quality; their strength. More important, if we are to achieve an effective vertical pathosystem management, we must only employ strong vertical genes.

A final comment is necessary concerning the vertical protection provided by pesticides. If the breakdown of the protection occurs quickly and the matching vertical pathotype disappears slowly, the vertical protection is weak. Conversely, a vertical protection may be strong and it should then be possible to employ some form of pattern (4.4) of strong vertical protection chemicals. However, Keiding (1967) has summarised evidence indicating that many protective chemicals must be classified as weak.

4.3.3 The Natural Pathosystem

We must now consider the function of weak and strong vertical genes in the natural pathosystem, and, because natural pathosystems have hardly been studied at all, we are compelled to use theoretical arguments only. There are two points.

Why should weak genes occur if their survival value is apparently less than that of strong genes? Let us consider a model of a natural pathosystem in which there are many vertical genomes of equal strength. If all the vertical genomes are weak, the exodemic is only slightly reduced; if all the vertical genomes are strong, the exodemic is greatly reduced. Let us suppose that all the vertical genomes are so strong that a matching allo-infection is a very rare event. That event could be so rare that the selection pressure for horizontal resistance would be dangerously low. There would be a loss of horizontal resistance and the esodemic would become too destructive, even if it was a rare event. In other words, there is an optimum strength of vertical genes, determined by the optimum levels of horizontal resistance.

Why should there be any difference at all in the strength of vertical genomes in a natural pathosystem? Let us reconsider the model of a natural pathosystem with the difference that there is wide variation in the strength of vertical genomes. The weak genes would tend to become extinct; the strong genes to dominate. There would then be a loss of host population heterogeneity which, as we have seen, is essential for the permanence of the vertical pathosystem. If one strong genome predominated, the esodemic would occur at a higher systems level; it would involve the population, not the individual. The vertical pathosystem would have lost balance. In other words, it seems that in a natural pathosystem, there is an optimum strength of vertical genome and that all vertical genomes are close to that optimum.

A natural pathosystem must be balanced. There must be a balance between the vertical and horizontal pathosystems. If vertical genes are too strong, the autonomous control of the esodemic is weakened, due to a loss of horizontal resistance. If the strength of vertical genes varies too far from the optimum, the autonomous control of the exodemic is weakened due to a loss of host heterogeneity.

4.3.4 The Crop Pathosystem

The crop pathosystem is vastly different. In the first place, cultivars differ from wild plants and modern cultivars, in particular, are the result of a deterministic, artificial selection, vertical genes having been collected, often from all over the world, for use in one locality. We can assume that all the genes possessed optimum strength in their native environments, but when brought by man to a new and different environment, some genes proved much weaker and others proved stronger than in their native conditions.

Secondly, cultivation differs from a natural ecosystem, involving in particular a deterministic control which permits the achievement of predetermined objectives. Some pathosystem characteristics are essential for an autonomous control; these include an optimum strength of vertical genes and a uniformity of vertical gene strength which is close to that optimum. Provided that the deterministic control is continuous and efficient, these characteristics are no longer essential and can be abandoned. We can artificially maintain high levels of horizontal resistance which are above the natural optimum (9.2) and, at the same time, employ vertical genes and combinations of vertical genes which are far stronger than the natural optimum.

A new picture begins to emerge. Balance in the natural pathosystem is represented by a level of disease such that the evolutionary survival of the host is not impaired. We call this the natural level of disease or damage. It now seems that, in the crop pathosystem, we can both employ and maintain levels of both horizontal and vertical resistance which are artificially high. If it is properly managed, the crop pathosystem would have less disease and damage than the natural level; we can even contemplate the perhaps remote possibility that it has no parasitism whatever.

4.3.5 Back-crossing without a Vertifolia Effect.

We have seen that horizontal resistance can only be measured and demonstrated in the esodemic; that is, after vertical resistance has broken down. The vertifolia effect is a host erosion (6.4) of horizontal resistance which occurs during breeding for vertical resistance. It occurs because the screening for vertical resistance is conducted during the esodemic with non-parasitic populations of the parasite; that is, with non-matching vertical pathotypes. This is the equivalent of breeding in the complete absence of the parasite; there is no selection pressure for horizontal resistance and, after some 10–15 host generations, the horizontal resistance erodes to the unselected level (6.4) which is a high agricultural susceptibility.

When breeding for vertical resistance, we really have no choice. We must work with parasitic populations of the parasite; that is, with vertical resistances which have already broken down. This is because we can only measure the strength of a vertical gene or genome by studying the matching vertical pathotype, and unless we are working with strong vertical genes, we are wasting our time. Equally important, we can only avoid a vertifolia effect by measuring the horizontal resistance in each generation of a back-crossing programme which consequently requires both matching and non-matching vertical pathotypes. All the host individuals of each generation must first be screened with the appropriate non-matching vertical pathotype; those individuals which are susceptible lack the vertical gene (or genome) and are discarded. The surviving individuals must then be screened with the matching vertical pathotype to determine their levels of horizontal resistance which can be expected to show a normal distribution. Only those individuals with the highest levels of horizontal resistance are retained for use as parents in the next generation.

4.3.6 The Domestication of Vertical Resistance

We have seen that some 70 years of scientific breeding for disease resistance have been dominated by vertical resistance, which was largely fortuitous and whose results were largely disappointing. It is easy to be critical, but it now seems that the work on vertical resistance is at last being vindicated.

Breeding for vertical resistance involves finding vertical genes from many parts of the world and incorporating them in cultivars for use in one locality. Some of these genes prove to be weak and are discarded, others prove to be strong and are retained. Even more important, new combinations of genes are tested; some of them are weak but, gradually, stronger and stronger combinations are discovered. This process is domestication, which was initially unconcious and ill-informed, but is now becoming deliberate and well-informed with a corresponding reduction in time scales. Eventually, a Hegelian change may be reached so that the parasite loses epidemiological competence and is then no longer a parasite. Vertical resistance has then come into its own; there is a boom but there is no bust (4.5). We must now discuss practical aspects of how this can be achieved.

4.4 Patterns of Strong Genes

4.4.1 Measurement of Strength

If we are to utilise strong vertical genes in our vertical pathosystem management, we must first measure the strength of vertical genes, both singly and in combination. There are three approaches.

The first approach was devised by van der Plank (1968) and involves measuring the rate at which a vertical pathotype becomes rare after having been common. This can be assessed either from vertical pathotype surveys in which the vertical genome of the pathotype must be related to that of the pathodeme, or it can be assessed experimentally, either in field plots or in a glasshouse or growth chamber. The relative half-life of the vertical pathotype is determined under conditions in which its vertical genes represent an unnecessary survival value. The original description should be consulted for details.

The second approach is historical and, for this reason, of somewhat limited usefulness, being essentially an assessment of the rarity of a vertical pathotype before it becomes common. The time which elapses before a vertical resistance breakdown is a function of both the vertical mutability and the selection pressure in its favour. Vertical mutability governs the appearance of the pathotype; selection pressure governs its population increase after it has appeared. If it is available, historical evidence of a long period of unbroken vertical resistance, in spite of high selection pressure, is thus evidence of strength.

The third approach is indirect and involves a determination of the reduction in epidemiological competence which is associated with the vertical genome in question. This assessment may be either incomplete or of limited value because the factors detracting from epidemiological competence are usually unknown, but useful circumstantial evidence can often be quickly and easily obtained from comparative horizontal pathogenicities, spore germination and longevity rates, etc.

4.4.2 Exodemic Manipulation

There are four factors of cardinal importance in all management of the vertical pathosystem. (1) Vertical resistance can only reduce the exodemic. (2) Although our control of the host population is direct, our control of the parasite population is indirect and is achieved by manipulating the host population for the maximum disadvantage of the parasite. (3) By the proper employment of deterministic control, we are relieved of the necessity of optimum strength of vertical genes and can exploit genes which confer the maximum strength. (4) If we can achieve a Hegelian change in the reduction of epidemiological competence, the first three of these factors become redundant (4.5).

In practice, this means that we must use the strongest available vertical genes and genomes and must enhance the role of exodemic by using them in properly designed patterns. At this stage, there is surely no need to re-emphasise the importance of the system, as a system; or the importance of the pattern as the most important fundamental in the systems concept. The patterns of vertical resistances may be spatial or sequential or both.

4.4.3 Spatial Patterns of Plants

A spatial pattern of plants is a multiline (Jenson, 1952) which provides a spatial discontinuity of host tissue. A good multiline has three characteristics; it has genotypic diversity, with respect to vertical genes, and phenotypic uniformity with respect to all agriculturally important characters. Each vertical genotype within the multiline must be strong to ensure that the reduction of the exodemic is maximal. Lastly, each vertical pathodeme must have a high horizontal resistance to ensure that the reduction of the plant-esodemic is maximal. If these conditions are met, the effects of vertical resistance will be quantitative and permanent. The complex vertical pathotype which matches all the vertical pathodemes in the multiline may occur but it will be rare. The combined reduction of both the exodemic and esodemic is a reduction of the epidemic.

The permanence of the vertical resistance depends on heterogeneity of the host population being maintained. In practice, this can be done either deterministically, by the mechanical mixing of seed of pure lines to produce the multiline, or autonomously. Provided the vertical genomes are of approximately equal strength, the proportions of the different vertical pathodemes will remain approximately equal.

Even the best multiline has two disadvantages; it is extremely difficult to produce a multiline with vertical resistances to more than one parasite. Sound pathosystem management must aim at comprehensive resistance to all the locally important parasites. A multiline should accordingly be used as a last resort, against that last and most difficult parasite of the pathosystem which, perhaps, can be controlled in no other way.

This, however, emphasises the second disadvantage, which is that there may be no vertical resistance to the parasite in question. Or, if vertical resistance does occur, the gene strength and/or the numbers of strong genes may be inadequate. Perhaps these comments reveal more than most, both the overall limitations of vertical resistance and the extent to which it has been over-emphasised in the past. Multilines may occasionally prove valuable, but these occasions will be relatively few.

4.4.4 Multiline Effects

We can now calculate the effects of a multiline. For simplicity, we shall assume that every vertical pathodeme in each multiline possesses only one vertical gene and that all vertical pathodemes occur in equal proportions. A multiline of five vertical pathodemes would be represented by the segment H2–H6 of Figure 7. We can formulate corresponding models for multilines with any number of monogenic pathodemes and calculate the probability of successful allo-infection with various, theoretical parasite populations. (Note that all the following models ignore the possibility of the vertical pathotype with no vertical genes).

If the parasite population consists of monogenic vertical pathotypes only, in equal proportions, the probability of successful allo-infection is closely similar to that of the natural pathosystem (Fig. 9) and is shown in Figure 10(1). This is a situation we would expect with strong vertical genes. The equal proportions of the pathotypes will be maintained throughout the exodemic if all vertical pathodemes have the same horizontal resistance. If the parasite population consist of both monogenic and digenic pathotypes, in equal proportions, the probability of successful allo-infection is higher and is also shown in Figure 10(2). This situation becomes increasingly less probable with increasing strength of vertical genes. If the parasite population consists of monogenic, digenic and trigenic pathotypes, in

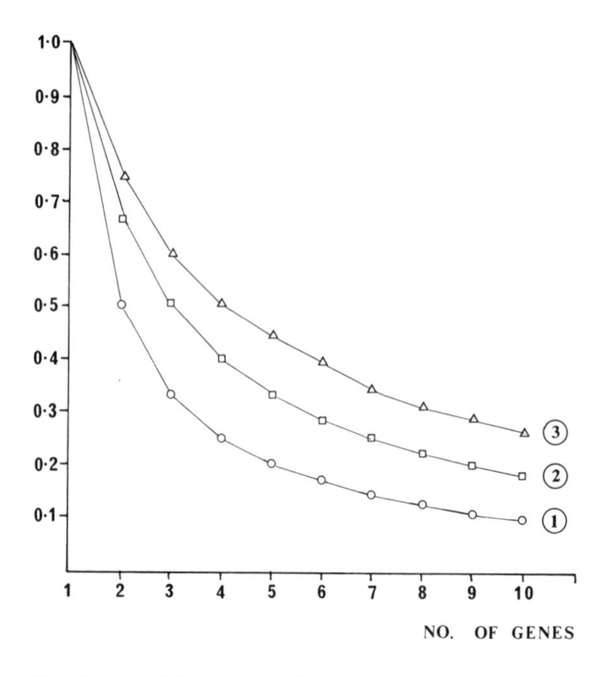

Fig. 10. Probability of matching allo-infection in multilines of monogenic pathodemes

equal proportions, the probability of successful allo-infection is again higher and is also shown in Figure 10(3). This situation becomes even more improbable with increasing strength of vertical genes. Clearly, all combinations of pathotypes are possible and the extreme is a uniform parasite population consisting of the "super race" which is the one complex pathotype which matches all the monogenic pathodemes of the multiline. The probability of successful allo-infection is then 1.0. But the occurrence of the "super race" would depend on the use of weak vertical genes and would become increasingly improbable with both an increasing gene strength and an increasing number of vertical genes in the multiline of monogenic vertical pathodemes. Three final comments are necessary. First we have assumed that all the vertical pathodemes of the multiline have an equal horizontal resistance. The level of the horizontal resistance will determine the total reproduction of a polycyclic parasite and, hence, the number of allo-infections within the crop (assuming no propagules coming from outside that crop) but the proportion of successful allo-infections will remain constant. Secondly, for simplicity, we also assume that each vertical gene in the multiline of monogenic vertical pathodemes has an equal strength. The level of this gene strength will determine the frequency of over-lapping vertical pathotypes, and, hence, the probability of successful allo-infection (Fig. 9). No doubt, many different computer simulations could be made with varying gene strength and varying horizontal resistance. Lastly, Figure 10 gives an indication of the increased size of parasite population required for the saturation technique for eliminating the effects of vertical resistance during screening for horizontal resistance (p. 102).

4.4.5 *Spatial Patterns of Crops*

A spatial pattern of crops is the same pattern as a multiline but it is at a higher systems level. The unit of pattern, or the individual of the population, is the crop. Each crop is the equivalent of an individual coffee tree; each plant in the crop is the equivalent of an individual coffee leaf. A spatial pattern of crops thus aims at the reduction of the crop exodemic. A crop pattern can be designed in one of two ways.

There may be a seasonal movement of the parasite, which may be from low latitude to high latitude with advancing summer; or a cyclical movement back and forth across the equator, governed by the movement of the inter-tropic convergence zone which also governs rainfall. The spatial pattern of crops should then be latitudinal so that, at regular intervals during the parasite migration, there is a different and strong vertical resistance to be overcome.

Alternatively, the crop-exodemic can be reduced within one climatic zone by a more or less random checker-board pattern. In general no crop can allo–infect its neighbours or be allo-infected by its neighbours.

4.4.6 *Sequential Patterns of Crops*

A sequential pattern of crops means that, at any one time, only one vertical pathodeme is cultivated but that, at regular time intervals in the pathosystem history, there is a complete change of pathodeme. There are three categories of sequential pattern. The first involves the individual field in the individual farm. This is rotation, with the difference that it is a rotation of different vertical pathodemes of one crop species, and is of value primarily against parasites of low dissemination efficiency such as Fusarium and Verticillium wilts and various nematodes.

The second is a regional pattern within one season. For example, in the higher latitudes, potato cultivation is divided into early and main-crop potatoes. Vertical genes 1, 2, and 3 against *P. infestans* are strong genes. It might be feasible to enforce legislation which prohibits the cultivation of any early potato variety possessing one or more of these genes. Except for the occasional tuber carrying blight the vertical resistances of the main crop potatoes would then recover each season. The blight epidemic would become two separate epidemics, each of entirely different vertical pathotypes. It would be interesting to see this attempted, at least experimentally, in a pathosystem with well-defined geographical boundaries, such as the United Kingdom or New Zealand.

The third category of pattern is also regional but is between seasons, being no less than a controlled boom-and-bust cycle. Provided that strong genes were employed and that all crops were carefully monitored, such a pattern could result in the advantages of a boom but avoid the disadvantages of a bust.

4.4.7 *Combinations of Patterns*

Combinations of more than one pattern are clearly possible but their use is likely to be limited. It must be remembered that we aim at comprehensive resistance. Vertical resistance does not occur against all parasites and the availability of adequately strong genes may be insufficient. It should be remembered also that patterns of vertical pathodemes are normally possible against only one parasite and that every increase in the complexity of pattern requires additional strong genes.

4.4.8 Complex Vertical Pathodemes

A pathodeme which possesses many different vertical genes is sometimes described as having "multiple resistance" or "broad-based resistance". These terms should be avoided because they refer equally to a multiplicity of parasite species and vertical pathotypes of one species.

A complex vertical resistance has several advantages. Principally, a complex genome is likely to be stronger than a simple genome, which means that a vertical genome with several, moderately strong genes may have the same overall strength as a genome with one, very strong gene. The total number of strong genomes, for use in patterns of different vertical resistances, can then be increased.

Secondly, if an inverse correlation between vertical and horizontal pathogenicities can be demonstrated, it would follow that a complex vertical pathodeme can only be infected by the equally complex vertical pathotype which would have a reduced horizontal pathogenicity. A reduced horizontal pathogenicity has the same effect as an increased horizontal resistance; there is less disease. This is an apparent horizontal resistance and the effect is permanent.

Lastly, as we have seen, a small increase in vertical genome strength may lead to a Hegelian change in the parasite survival. This topic must now be discussed in some detail.

4.5 The Hegelian Change in Strength

4.5.1 Description

Van der Plank (1975) has discussed the genome Sr_6 Sr_{9d} for vertical resistance to *P. graminis* in the spring wheats of North America, on the basis of papers by Green (1971) and Stewart *et al.* (1970). There seems to be little doubt that this vertical genome is so strong in Canada that the matching vertical pathotype lacks epidemiological competence. It is common in Texas and vertical pathotypes with either one of these two genes are reasonably common in Canada. This vertical resistance has remained effective in Canada for some twenty five years, in spite of minor epidemics of matching pathotypes in 1952 and 1960. Normally, such a minor epidemic heralds the "bust" of a vertical genome. However, for the first time, the boom has continued and, it seems, the bust cannot happen.

The explanation has already been mentioned. It is believed to be a Hegelian change in vertical genome strength. With a small increase in the degree of strength, the ability of the parasite to survive has been lost completely; effectively, it is no longer a parasite, in that area, on cultivars with that genome and in a normal season.

4.5.2 Definitions

The first question arising from the above explanation concerns the effect of this new situation on our existing definitions; do they have to be modified? Let us consider the nature of a Hegelian change. When we define the properties of water, we are describing a substance which, by definition, is a liquid. If there is a Hegelian change and the water becomes ice, our definitions of water are not affected, although they do not apply to the solid ice. It is the same with vertical resistance. When the Hegelian change occurs, the resistance is no longer vertical resistance. We might, perhaps, call it *frozen* vertical resistance. An analogy might be useful. Given the Hegelian change in temperature, precipitation becomes frozen and solid, and an ice cap forms. In an abnormal summer, some ice may melt, but the ice cap is permanent, at least for the foreseeable cultural future. Given the Hegelian change in vertical genome strength, vertical resistance "freezes". In an abnormal summer it may "unfreeze" to some extent, but the general state is permanent, at least in the foreseeable agricultural future.

Frozen vertical resistance has many of the behavioural characteristics of horizontal resistance; it is an apparent horizontal resistance, but definitely not true horizontal resistance, which has a fundamentally different structure at all system levels. Again, by analogy, we could comment that ethyl alcohol has many of the characteristics of water; it is an apparent water, but it is not true water because it has a fundamentally different structure.

4.5.3 The Natural Pathosystem

The second question concerns the natural pathosystem. Does frozen vertical resistance occur in a natural pathosystem and, if not, why not?

It is indeed tempting to suggest that frozen vertical resistance is normal in a natural pathosystem and that, in the course of both cultivation in different environments, and domestication, man has unwittingly reversed the Hegelian change and unfrozen many vertical resistances. This would explain many of our crop pathosystem troubles, but the suggestion is inherently unlikely. In the course of fairly primitive cultivation and domestication, the unfreezing of vertical resistance would be a disaster of such magnitude that the primitive system would probably collapse completely, and there is no evidence of this having occurred. Evidence from many thousands of years ago, is, of course, lost but we must also consider the world-wide distribution of crop species which has occurred during the past few centuries. The somewhat alarming possibility that we may inadvertently unfreeze a currently frozen vertical resistance can accordingly be dismissed.

It seems, therefore, that frozen vertical resistance does not occur in natural pathosystems and we must enquire why not. In the first place there is the general, quite new evolutionary consideration that man himself is a Hegelian change. His deterministic control of his own ecosystem and of much of evolution itself has led to many other Hegelian changes, all of which are quite new to the evolutionary system. The freezing of vertical resistance is not remarkable just because it is new.

Moreover, the freezing of vertical resistance was achieved quite fortuitously in the course of a somewhat random process of trial and error, a process which resulted in extreme losses of systems balance and, inevitably, was remarkably destructive. As we have seen, the loss of balance was two-fold. There was a loss of host population heterogeneity to the point where the control of the exodemic by vertical resistance was lost, and also a loss of horizontal resistance to the point where control of the esodemic was also lost; hence the boom-and-bust cycle. Such a process is clearly possible in a crop pathosystem in which man's attempts at pathosystem management are in a relatively early, primitive, ignorant and clumsy stage of development. Such a process must surely be impossible in a natural pathosystem in which systems balance, both between host and parasite and between all the hosts of the ecosystem, is essential. It has been argued that surviving, natural pathosystems can only be balanced, otherwise they would not have survived, and this is obviously true at every interval in the evolutionary systems history. It has also been argued that the requirements of autonomous control are an optimum strength of vertical genes and that all the vertical genes of a pathosystem must tend to this optimum, which is in accordance with all principles of homeostasis. Some variation occurs because the system must be resilient; it must be able to recover from quite wide swings away from the optimum balance. But the system must also be stable and these swings must be prevented as far as possible. It is extremely doubtful if a natural pathosystem could survive a boom-and-bust cycle of the intensity we have witnessed in some crop pathosystems during the past half century.

4.5.4 The Crop Pathosystem

We must now enquire how frozen vertical resistance can be utilised in crop pathosystem management. There seems to be no doubt that frozen vertical resistance exists in these spring wheats of North America. The obvious first question is whether or not the frozen vertical resistance in these spring wheats is permanent. If it is to prove impermanent, this may be for one of two possible reasons. There is only one vertical pathotype which exactly matches the frozen vertical resistance; all other pathotypes which can allo-infect it successfully are over-lapping vertical pathotypes. That is, their loss of survival ability should be even greater than that of the exactly matching pathotype. It is possible that the exactly matching pathotype has not yet appeared and that it will "unfreeze" the vertical resistance when it does appear. (Alternatively, a "new" overlapping pathotype might do this for reasons unknown). An existing pathotype might also "unfreeze" the vertical resistance by changing to the extent that it gains the same vertical pathogenicity by the loss of other, different survival values. Clearly, we cannot resolve these possibilities but they seem unlikely.

The second question is whether or not the existing, frozen vertical resistance can be improved. Obviously it can, in various ways. By analogy, frozen water can be more deeply frozen, which does not change its state but does make permanent ice less liable to melt in an abnormal summer. Secondly, the various cultivars of the frozen vertical pathodeme can be improved with respect to their horizontal resistance to stem rust their comprehensive resistance to all locally important parasites and all other improvements which result from cumulative plant breeding (9.2). The third question concerns wheat stem rust in other parts of the world. Can we freeze its vertical resistance in these areas also? Here the implication seems to be that variations in vertical gene strength increase with increasing environmental difference from the natural pathosystem; that is, extremes of gene strength will most likely occur at the furthest climatic distance from the centre of origin of the crop species. Wheat has a polyphyletic origin and, as a result, it has several centres of origin and diversification; and *P. graminis tritici* can only be a polyphyletic hybrid (Chapt. 7). It seems, therefore, that we can freeze this vertical resistance in other areas, particularly if we use combinations of vertical genes which are far removed from the environment of their natural pathosystems.

The fourth question is whether or not we can freeze vertical resistance in other crop species. Presumably we can, although the ease of doing so is likely to vary greatly with different crops and parasites, and may sometimes be impossible.

The fifth and most important question is how we go about it. Here the most important consideration is the knowledge that it can be done, however difficult this may or may not prove in practice. The process must obviously be one of domestication, a gradual increasing in the strength of the known vertical genomes until the Hegelian change is reached. As we have seen, domestication is much faster when it is both deliberate and well-informed. In the past, breeding for vertical resistance consisted of replacing old genes with new and the domestication of strength was largely random, but now we are able to measure the strength of vertical genes and genomes. No doubt, techniques for doing this will become easier and more precise.

The final question concerns the limitations of frozen vertical resistance. Clearly, its permanence is an obvious advance over the boom-and-bust cycle, but it still operates against only one parasite of a pathosystem. The achievement of one frozen vertical resistance will be fairly difficult, and to achieve frozen vertical resistance to two parasites in one cultivar may well be impossible. If we are to achieve comprehensive resistance to all the important parasites of a crop pathosystem we must use horizontal resistance also.

Chapter 5 Horizontal Pathosystem Analysis

5.1 General

5.1.1 Recapitulation

The term horizontal is derived from the diagram (Fig. 2) of van der Plank (1963) Agriculturally, horizontal resistance is permanent resistance. Its use does not lead to a boom-and-bust cycle of cultivar production and breeding for horizontal resistance should be cumulative, a good cultivar being replaced only by a better cultivar. A useful general rule is that horizontal resistance provides an incomplete but permanent disease control. Polygenically-inherited resistance is always horizontal resistance but not all horizontal resistance is inherited polygenically. Horizontal resistance is indicated by the absence of a differential interaction in the amounts of disease when a series of different pathodemes is inoculated with a series of different pathotypes. That is, there is a constant ranking of pathodemes, according to resistance, regardless of which pathotype they are tested against; and likewise a constant ranking of pathotypes, according to parasitic ability, regardless of which pathodeme they are tested against. Horizontal resistance is conferred by mechanisms which are beyond the parasite's capacity for change; horizontal resistance and horizontal parasitic ability are independent of each other.

Because vertical resistance is valueless in the esodemic, it follows that horizontal resistance is essential. Because every epidemic has an esodemic it also follows that horizontal resistance is universal, occurring in all plants against all parasites, even though it may be agriculturally inadequate in many cultivars. The esodemic involves the spread of the parasite within genetically identical host tissue, which may be the individual leaves of one tree or the individual plants of one clonal or pureline crop. The esodemic begins with the breakdown of vertical resistance, if it is present. In vertically resistant crops, horizontal resistance can only be measured and selected in the presence of a matching vertical pathotype. The natural level of horizontal resistance may be inadequate to control a disease in crops with vertical resistance.

5.1.2 Oligogenic Horizontal Resistance

Van der Plank (1963, 1968) has reported oligogenic horizontal resistance in three diseases: loose smut of barley *(Ustilago nuda hordi)*, onion smudge *(Colletotrichum circinans)* and milo disease of sorghum *(Periconia circinata)*. It seems that horizontal resistance to bacterial blight *(Xanthomonas malvacearum)* in cotton can also be inherited oligogenically. Resistance to woolly aphid *(Eriosoma lanigermum)* in apple is inherited oligogenically and is horizontal; it was first recorded in 1831 (Painter, 1951).

Oligogenic horizontal resistance is rare. Anyone interested in finding other examples should perhaps, look for oligogenically inherited resistance either to insect, bacterial and virus parasites or in crops whose progenitors grew in uniform populations. When it occurs, oligogenic horizontal resistance is very valuable, as it combines the main advantages of vertical and horizontal resistance, and the easy breeding technique and qualitative effects with permanence. But its rarity makes it of minor interest except in a terminological context; not all oligogenic resistance is vertical resistance and not all horizontal resistance is inherited polygenically. Oligogenic horizontal resistance is qualitative in its inheritance, its mechanisms and its effects and, for this reason, it can be confused with vertical resistance. Unless stated to the contrary, all discussion in this book concerns polygenic horizontal resistance which is quantitative in its inheritance, its mechanisms and its effects.

5.1.3 Apparent Horizontal Resistance

We have seen that some indications of the vertical nature of resistance are false. The same is true of horizontal resistance; the effects of vertical resistance can occasionally be mistaken for horizontal resistance. This is most common when those effects are quantitative, as with incomplete vertical resistance or multilines; false syllogisms have already appeared in the scientific literature. Equally, a complex vertical pathodeme can only be parasitised by the matching vertical pathotype which is complex and consequently may have a reduced horizontal parasitic ability. Frozen vertical resistance is also an apparent horizontal resistance.

5.1.4 Immunity

It has already been suggested that immunity should be regarded as being outside the conceptual boundaries of the pathosystem. The possibility of absolute horizontal resistance is discussed on p. 94 and this has all the characteristics of immunity except one. Immunity will not erode in the absence of the parasite while absolute horizontal resistance will do so. Immunity is a non-variable survival value; absolute horizontal resistance is the extreme of a variable survival value.

5.1.5 Tolerance

The word tolerance has been used to mean so many things that special comment is necessary. We must first consider what tolerance really is; we must then examine it in relation to resistance and, finally, we must discuss its relationship to horizontal resistance.

Tolerance is usually taken to mean that, when two cultivars are equally diseased, the more tolerant one suffers a smaller loss of yield. To understand tolerance, we must recognise that there are three ways of measuring disease (or other parasite damage). These are: the amount of visible symptoms; the amount of reproduction in the parasite; and the amount of yield loss in the host. (In very general terms, yield of the host is equivalent to reproduction of the host). Normally, the three different measures coincide closely. However, if there is a relatively high visible disease, or a relatively high parasite reproduction, combined with a relatively low yield loss, the host is described as tolerant. But the term should be used with caution because tolerance is difficult to demonstrate conclusively. For example, two cultivars may have widely different yields when disease-free, and similar yield differences are to be expected when they are equally diseased. However, let us assume that tolerance has been conclusively demonstrated.

The growth of a parasite is a physiological sink in the host and, at the same time, it either destroys or interferes with a physiological source in the roots, stems or leaves. Given equal visible disease or equal pathogen reproduction, there are only two possible explanations for differences in yield. The more tolerant host may possess mechanisms which lessen the effects of these physiological sinks and reduced physiological sources, but this is inherently unlikely. Alternatively, the less tolerant host may possess mechanisms which exaggerate the effects of the physiological sinks and reduced physiological sources. There may also be secondary effects such as an exaggerated sensitivity to metabolic byproducts of the parasite. In other words, the tolerant host suffers a normal loss of yield, and the less tolerant host an abnormal loss of yield.

The implication is clear. Tolerance is normal and natural; a loss of tolerance is abnormal and artificial. Tolerance represents systems balance; a loss of tolerance represents a loss of systems balance. If we breed for tolerance, we are merely restoring systems balance which, presumably, was lost in the process of domestication. We must also enquire how tolerance is related to resistance. In so far as resistance is normal and natural in a wild pathosystem, susceptibility is abnormal and artificial, being a loss of systems balance which occurred during domestication. At the epidemiological level, a loss of tolerance must be regarded as the same as a loss of resistance. At the histological or physiological level of the pathosystem, we recognise that a loss of tolerance is due to an abnormally increased sensitivity; the loss of resistance is due to an abnormally reduced efficiency of resistance mechanisms. At the epidemiological level of pathosystem analysis, however, the effects are identical. Epidemiologically, tolerance, like disease escape, is a component of resistance.

If, in epidemiological terms, tolerance is the same as resistance, is it vertical or horizontal resistance? There seems little doubt that a high level of tolerance is beyond the parasite's capacity for change, nor is it likely to lead to a boom-and-bust cycle of cultivar production. We must consequently regard tolerance as a component of horizontal resistance. However, not all horizontal resistance is due to tolerance, and again there is a danger of false syllogisms. Some authors consider the term tolerance to be synonymous with horizontal resistance, but this is clearly incorrect.

Finally, two special cases must be noted. Hybrid maize carrying Texas, cytoplasmic male sterility is abnormally susceptible to Race T of *Helminthosporium maydis* and oats carrying the gene Pc-2, which confers vertical resistance to *Puccinia coronata*, is abnormally susceptible to *H. victoriae*. In each case the susceptibility is due to an extreme sensitivity to a mycotoxin. It is tempting, therefore, to regard other cultivars of the same host species which are unaffected by these toxins as tolerant, and to equate this tolerance with a differential interaction, a single inheritance factor and, hence, vertical resistance. However, vertical resistance is due to mechanisms which are within the capacity of change of the parasite, which means that the mechanism is present but that it may fail to function. These two special cases are different. The single inheritance factor does not confer a resistance mechanism, but an abnormal loss of tolerance, this tolerance is present and normal in all other cultivars.

This is obviously a very unusual and a very unnatural situation and it illustrates the extreme of abnormality which can occur in a crop pathosystem. It is also a very rare situation and its significance has been grossly exaggerated in the scientific literature.

5.1.6 *Disease Escape*

A cultivar which matures early is harvested early, often before the epidemic has reached damaging proportions, and this phenomenon is known as disease escape. At the pathosystem level of resistance mechanisms, disease escape is not a resistance mechanism, but at the epidemiological level, it is a factor contributing to the reduction of disease, and, in general terms, a form of resistance. It is also manifestly beyond the capacity of micro-evolutionary change of the parasite and is thus a component of horizontal resistance.

5.1.7 *The Pathosystem Level*

It was commented earlier that one of the advantages of an abstract term such as horizontal, is that it can be used with equal precision at any systems level.

At the level of the host individual, horizontal resistance is conferred by many different mechanisms; it consists of many different and complex, variable survival values. In the cereals, for example, these may include adult plant resistance, slow rusting, the Hart (1931) phenomenon (in which a high proportion of sclerenchyma and a thick cuticle restrict the production and liberation of uredospores respectively), tolerance and disease escape. At this pathosystem level, some of these mechanisms are not even resistance, in the strict sense of the term. Collectively, these mechanisms, by reducing the rates of infection and colonisation of the host and reproduction of the parasite, reduce the effects of parasitism after the host-parasite contact has been made, and thus reduce the esodemic.

At the crop population or epidemiological level, however, horizontal resistance is not necessarily due to any one of these mechanisms, but is a single survival value which is usually quantitative in its inheritance and effects.

5.2 Horizontal Resistance Is Universal

In the past, work on vertical resistance so dominated both breeding and pathology that many workers in these fields began to conclude that all resistance was vertical and that any form of permanence in their work was impossible. In this climate of opinion, which still exists, any suggestion that horizontal resistance not only occurs, but that it occurs in all plants against all parasites, is likely to be received with scepticism if not with scorn. Nevertheless, all the evidence indicates quite clearly that vertical resistance is an exceptional and supplementary form of resistance in nature, and that horizontal resistance does indeed occur in every plant, against every parasite, even if, in many cultivars, it currently occurs at an agriculturally inadequate level. This postulation is so contradictory to popular opinion and so important to agriculture that evidence for it must be examined; it falls under six headings as follows.

5.2.1 *Absolute Susceptibility*

Van der Plank (1968) first pointed out that it was unlikely that vertical resistance ever occurred without horizontal resistance. Horizontal resistance is the resistance which invariably remains after vertical resistance has broken down. Many cultivars are very parasite-susceptible following the breakdown of vertical resistance but they still have some resistance; otherwise they would exhibit absolute susceptibility. This is a condition in which a parasite, and particularly a pathogen, could grow through living host tissues at its maximum rate of growth without any impediment whatever. It is comparable to the growth of a facultative parasite on the optimum nutrient agar. The nearest approach to absolute susceptibility probably occurs in senescent, expendable host tissue such as an over-ripe fruit. But absolute susceptibilty never occurs in living host tissues and, therefore horizontal resistance is present, even if only at a low level.

5.2.2 *The Esodemic*

Perhaps the best evidence comes from consideration of the esodemic. Vertical resistance is valueless against auto-infection and horizontal resistance is essential in the esodemic. Let us consider a theoretical model of a natural pathosystem in which there is wide range of different vertical resistances but no horizontal resistance whatever. We can assume that the mixture of vertical resistances is so effective that it reduces the frequency of successful allo-infection to the minimum, but it cannot prevent successful allo-infection entirely and some matching will always occur. The allo-infected individual would then exhibit absolute susceptibility and would be totally destroyed by the parasite. This would be the equivalent of the predator–prey relationship in which carnivores entirely consume relatively few individuals of a herbivore population. A natural population balance is maintained in a predator–prey relationship and such a relationship is theoretically possible in a plant pathosystem, but has never been observed.

5.2.3 *Domestication*

Cultivars differ from their wild progenitors in relatively minor ways. A few selection pressures were changed during domestication and a few survival values have been emphasised at the expense of others. However hard we try, we cannot make wheat susceptible to coffee rust, potato blight or any of many thousands of other plant parasites, nor to thousands of other micro-organisms. We tend to forget this in our concern over a few major parasites, which are only important because the parasitic relationship is unbalanced. Variable horizontal resistance has been allowed to fall below its natural optimum due to host erosion (6.4). This optimum can be restored by breeding. It can even be exceeded until it approaches absolute resistance.

5.2.4 Uniform Host Populations

The obvious trends of an advanced agriculture are towards an increased specialisation, an increased intensity of production and an increased uniformity of product. This is seen in wheat, for example, where the best wheat lands now grow wheat exclusively, in place of the older mixed farming; where yields and quality are raised to high levels and where bulk harvesting, storage and handling, as well as mass-markets, all require a high uniformity of product. This leads to increasing levels of monoculture, both in total area and sequence of crop, and also with respect to a single cultivar rather than a single crop species. Such monoculture can lead to ideal epidemic conditions with its high host population density and uniformity, but it is not necessarily unnatural and wild plants can occur in comparably uniform populations. Coniferous and bamboo forests, some prairies and savannas, hillsides covered in bracken, the papyrus of the Nile sudd and tracts of elephant grass are typical examples of wild plant populations which do not become parasitised beyond the natural level. Their evolutionary survival is not impaired by their parasites. Some of them, it should be noted, are clonal and have great genetic uniformity while others are long-term perennials and evergreens possessing no vertical resistances, and they could not survive without horizontal resistance.

The fact that natural monoculture occurs in some wild species does not necessarily mean that monoculture can be successfully practiced by man with any species he chooses to exploit; but it does indicate that an intensive, cultivated monoculture is possible in many crop species, depending on the nature of their wild progenitors.

5.2.5 Subsistence Cultivars

Much of the worlds' agriculture is conducted by peasant farmers on a subsistence or semi-subsistence basis. A subsistence crop is one grown by the farmer for the use of himself and his family, as opposed to a cash crop which he grows for sale and income. In so far as a peasant farmer spends his very limited money on the production of his crops, he devotes it to his cash crops because these will return his money with a profit, wheareas money spent on food crops is lost by being eaten. In some developing countries, as much as ninety per cent of the population may be engaged in farming on this basis. The internal market for food crops is then small and is usually supplied by subsistence crop surpluses. Food crops thus tend to be subsistence crops, while industrial crops, such as the beverages and fibres, tend to be cash crops.

Subsistence food crops are thus cultivated with the minimum of cash expenditure on parasite control. Some parasite control is achieved by traditional practices such as the burning of crop residues, shifting cultivation and the cultivation of both mixed stands of crop species and mixtures of varieties within a crop species. Although, collectively, these measures are not insignificant, they are trivial when compared with, say, the routine use of foliar fungicides. Nevertheless, severe parasite damage is generally rare in subsistance cultivars: they are resistant. Some of them, such as bananas and sweet potatoes, are clones, and others, such as rice, are self-pollinating lines of considerable antiquity and their levels of horizontal resistance are high. The clones in particular are propagated without any of the elaborate and expensive seed health certification that bedevils crops such as potatoes and strawberries in the more advanced agricultures.

5.2.6 *The Unselected Level of Horizontal Resistance*

Horizontal resistance is a variable survival value. In the presence of a parasite, there is positive selection pressure for it and horizontal resistance increases to the optimum necessary for the natural level of disease or damage. In the absence of the parasite, there is negative selection pressure and horizontal resistance decreases to its unselected level, a situation which occurred with the African maizes which were cultivated for some hundreds of generations in the absence of *P. polysora* (6.4). It is also the level of horizontal resistance reached when the vertifolia effect is at its maximum; when a genetically flexible host population is routinely treated with protective chemicals; or in a cultivar which is taken from an area where the parasite lacks epidemiological competence to an area where it has full epidemiological competence. We should note three points about the unselected level of horizontal resistance. (1) It can be determined experimentally by eliminating positive selection pressure for horizontal resistance in a genetically flexible population. (2) It is a higher level of resistance than the minimum horizontal resistance which can also be determined experimentally by an artificial selection for susceptibility. (3) The unselected level of horizontal resistance represents a high agricultural susceptibility (6.4). In other words, even the most susceptible cultivars still have some horizontal resistance, and this applies to all their parasites.

5.3 Horizontal Resistance Is Permanent

It has been postulated that horizontal resistance is permanent, at least during the foreseeable, agricultural fulture. This postulation is at least as important as the postulation that horizontal resistance is universal. Its discussion involves two categories of host population; the genetically flexible population, such as maize, and the genetically inflexible population such as a clone of potatoes or a pure line of wheat.

The term "permanent" must also be qualified, as it is the antithesis of the ephemeral effects of vertical resistance. We recognise that horizontal resistance is a variable survival value and that, under certain circumstances, its effectiveness may be reduced. One of the functions of pathosystem analysis is to ensure that this does not happen in pathosystem management. For all practical purposes, the resistance can then be described as permanent.

5.3.1 *The Natural Pathosystem*

The natural pathosystem is a dynamic system with autonomous control. Systems balance is maintained; it is a stable but also flexible system. As a result of changes in its environment, the system can exhibit quite wide swings away from the optimum and yet recover; it is a well-buffered, resilient system, which is why so many survival values must be variable; their variation provides resilience, homeostasis and stability.

Both horizontal resistance and horizontal parasitism are variable survival values. As such, they are subject to the mathematics of extremes; they have a minimum, an optimum and a maximum. In a natural pathosystem, there is systems balance; were this not so, the pathosystem could not have survived. This means that both the resistance and the parasitism tend to the optimum for the state of the system at that time. This optimum is a level of damage to the host such that the evolutionary survival of both the host, and the parasite with it, are not impaired. We call this the natural level of damage or disease, and it is produced by the interaction of the natural level of horizontal resistance and the natural level of horizontal parasitic ability.

5.3.2 The Crop Pathosystem

A crop is cultivated; that is, it is protected from much of the natural competition of a wild ecosystem. It is also domesticated; this means that certain variable survival values, such as yield and quality, have been increased while others, such as the ability to compete in a wild ecosystem, have been reduced. It is common knowledge both that cultivars are of far greater value to man than are wild plants, and that they cannot survive unaided in a wild ecosystem.

At one systems level, yield is a single survival value; one cultivar may have a higher yield than another. At a lower systems level, yield is itself a system; it is many patterns of patterns. It has many components each of which is a variable survival value, and each of which has a natural minimum, optimum and maximum. The inheritance of the survival values is usually controlled polygenically although oligogenic control does occur. In the course of domestication, most of these variables have been increased by artificial selection to a level above that of the natural optimum. The same is true of crop quality, and should also be true of resistance but is not. It is common knowledge that cultivars are more susceptible to parasites than are wild plants. At one systems level, resistance to all parasites is a single survival value. At a lower systems level, resistance has many components. There are separate resistances to many different parasites and the resistance to any one parasite involves many different mechanisms. Each mechanism is a variable survival value and its inheritance is usually controlled polygenically, although oligogenic control can occur. In the course of domestication, and in contrast to yield and quality, many of these variables have been reduced below the natural optimum; which could be called a negative domestication. Crop pathosystem management must involve a positive domestication, in which all of the variables contributing to resistance are increased, at least to the natural optimum and, preferably, to a level above the natural optimum. This can be done by artificial selection, that is, by plant breeding.

5.3.3 The Breakdown of Horizontal Resistance

With an extreme fluctuation in one or more environmental factors, horizontal resistance mechanisms may fail to operate, especially qualitative, oligogenically inherited horizontal resistance. Such a failure may be legitimately described as a breakdown of horizontal resistance, but is entirely different from a breakdown of vertical resistance, being due to a physiological change in the host and not to a change in the parasite population. Furthermore, it is a temporary failure, as with the return of normal environmental conditions, the resistance is again fully functional.

We should note this phenomenon for the sake of comprehensiveness and accuracy of definitions. Such a temporary failure of horizontal resistance does not affect its overall permanence, and we should note also that it is a rare phenomenon and that it is unlikely to be of practical significance in pathosystem management.

5.4 Genetically Flexible Crop Populations

Person (1975) produced a mathematical model to demonstrate the stability of the horizontal pathosystem when both the host and the parasite population are genetically flexible (Fig. 11). This is a density-dependent system in which autonomous control is exerted by negative feedback through genetic change. The genetic change consists of either a loss or a gain of horizontal resistance in the host population, and of horizontal parasitic ability in the parasite population. The mechanism of change is reproductivity. If the parasite loses parasitic ability, the host reproductivity increases, and, in the process, the host loses resistance due to negative selection pressure; the parasite reproductivity then increases at the expense of host reproductivity and the host gains resistance due to positive selection pressure. The host population exerts equivalent selection pressures on the parasite population. This is a special form of hunting in which the system as a whole fluctuates about the mean which, in Fig. 11, is represented by damage level 5. In a natural pathosystem, this level is the natural level of damage or disease; and it results from the interaction of the natural levels of both resistance and parasitic ability.

One point must be noted immediately. The variation in parasitic ability differs remarkably between parasite species. In a facultative parasite such as a Fusarium wilt, the variation is very wide and ranges from a near-obligate saprophyte to a near-obligate parasite, but in an obligate parasite, such as *P. infestans*, the variation in horizontal pathogenicity is very limited.

5.4.1 The Rossetto Hypothesis

The Rossetto (1975) hypothesis states that the non-uniform distribution of an insect parasite population can be a positive survival value for the parasite. This is because the host is unable to maintain high levels of horizontal resistance in the absence of continuous, positive selection pressure. A sequential non-uniformity is obvious and well known in a few insects such as locusts, of which there is only one major infestation in the course of some ten or so generations of an annual host and, as a result, the host cannot accumulate resistance. Rossetto has extended this

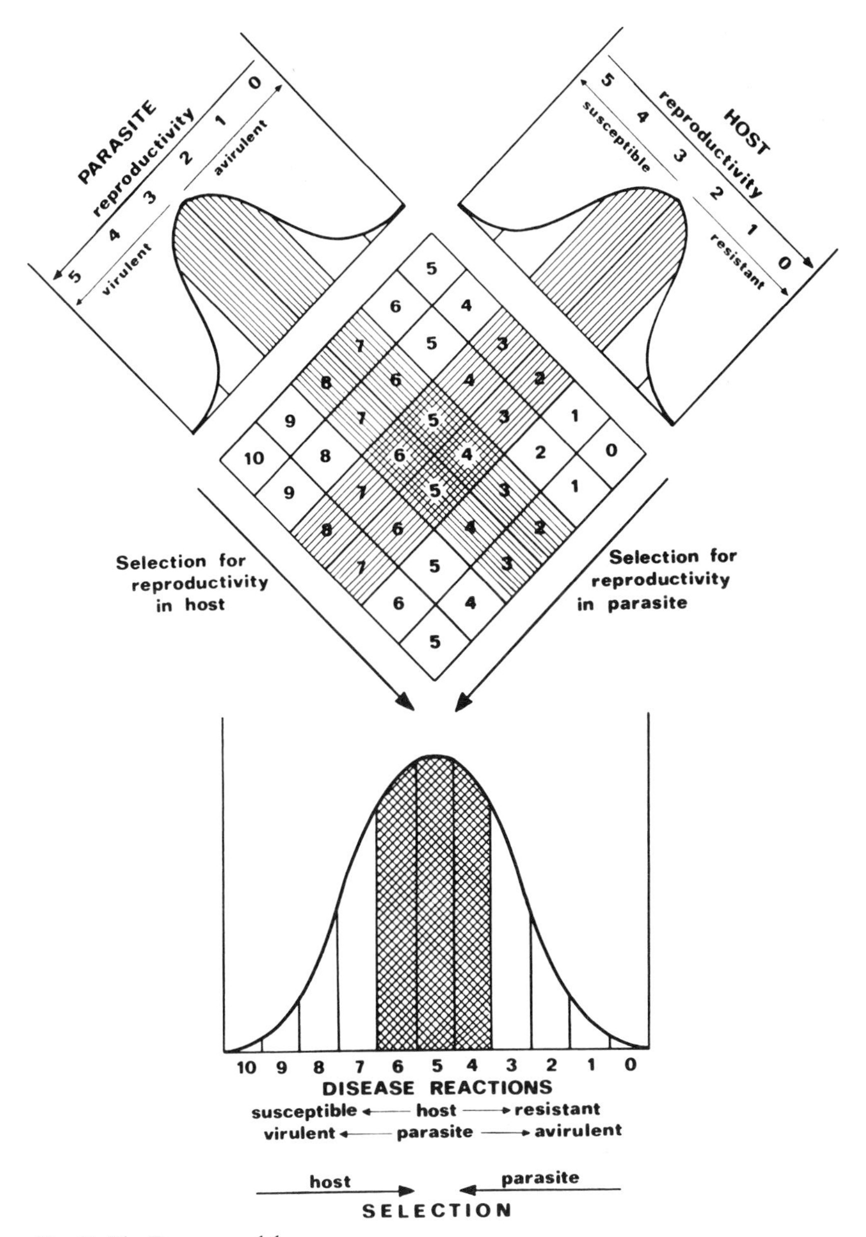

Fig. 11. The Person model

concept to spatial distribution. If a parasite is regularly absent from most of an annual host population each season, the host remains relatively susceptible. This situation approaches the theoretical model of an exclusively vertical pathosystem and the predator-prey relationship of carnivores and herbivores. The hypothesis also provides a mechanism for the "spotty distribution" of ecologists. Systems balance is maintained because the majority of host individuals escape entirely in spite of being susceptible. Obviously, this systems balance may be lost in a crop pathosystem where both spatial and sequential host population densities may be so unnatural that the normal restraints on parasite reproduction are removed. Rossetto has emphasised the necessity for artificially uniform parasite infestations during screening for resistance. If the infestation is natural and non-uniform, it becomes impossible to determine differences in levels of resistance.

5.4.2 *Maize Streak Virus*

Maize streak virus occurs in Africa and can occasionally be very damaging. Van der Plank (1963) suggested that the reason for this might be the fact that this is an indigenous virus attacking an exotic host. Later, Robinson (1973a) suggested that the severity of the disease was due to a regular host erosion of horizontal resistance which occurred because the disease was sporadic in both time and space. For this reason, an artificially selected high level of resistance could be eroded under peasant farming conditions (6.4). We now have a third, confirmatory explanation in the Rossetto hypothesis. This virus is transmitted by a leaf-hopper *(Cicadulina maydis)* and, presumably, the sporadic infestations represent a positive survival value for the vector.

5.4.3 *The Natural Level of Horizontal Resistance*

It is now necessary to define the natural level of horizontal resistance. This is the level of resistance which occurs in a natural pathosystem and is such that the parasite damage does not impair the evolutionary survival of the host. It is clear that with different pathosystems, this level can vary widely. It may be quite low under conditions of the Rossetto hypothesis, and may also be lower against a parasite where there is a vertical pathosystem than against a parasite where there is only a horizontal pathosystem. The natural level of horizontal resistance is thus related to parasite rarity, but against any given parasite in a balanced, natural pathosystem, there is a natural level of horizontal resistance, which could if necessary be determined experimentally.

5.4.4 *The Maximum Level of Horizontal Resistance*

It is clear that the natural level of horizontal resistance represents the optimum of a variable survival value and casual domestication of crops has usually led to a loss of horizontal resistance which is in adequate for the crop pathosystem. Deterministic domestication could lead to increases in horizontal resistance, both to a level above the natural optimum and to the maximum level, which could also, if necessary be determined experimentally.

5.5 Genetically Inflexible Crop Populations

We must now consider why horizontal resistance should be permanent in a genetically inflexible host population such as a clone of potatoes or a pure line of wheat or rice. The host population is artificial; it cannot change, it can only be discarded or replaced. The parasite population on the other hand is wild and natural; it is flexible and can respond to selection pressures. If it changes towards an increased horizontal pathogenicity, the effectiveness of the horizontal resistance will decline. This decline will be quantitative but, apart from this, it will be no different from the qualitative breakdown of vertical resistance. Both losses in the effectiveness of resistance would be due to changes in the parasite population. Horizontal resistance would then be temporary resistance and cultivar replacement would continue indefinitely. The time has come to demolish this persuasive but entirely false suggestion.

5.5.1 Constant Ranking

With traditional breeding for disease resistance, a vertical pathodeme becomes a cultivar for the simple reason that all locally known pathotypes are non-parasitic to it. Screening for vertical resistance is thus dependent on the very dubious criterion of non-parasitic populations of the parasite. Screening for horizontal resistance is very different, and, as we have seen, can only occur during the esodemic—that is after vertical resistance, if it is present, has broken down, and is thus dependent on parasitic populations of the parasite. The ranking of horizontal pathodemes, according to resistance, is constant, regardless of which pathotype they are tested against. In other words the resistance operates against all pathotypes, including the most parasitic ones. The most resistant pathodeme will always retain its rank, in spite of any changes in the parasite population. Breeding for horizontal resistance thus involves extending the series of pathodemes of Figure 4 in the direction of increased resistance. If we are to demonstrate the permanence of horizontal resistance, we must demonstrate that nature cannot extend the series of horizontal pathotypes in the direction of increased horizontal parasitic ability. That is, we must first of all demonstrate that there is a maximum horizontal parasite ability and, secondly, that all important parasites already possess that maximum.

5.5.2 The Artificial Level of Disease

We must now define the artificial level of damage or disease, which occurs in most crops. It is a level higher, and often very much higher, than the natural level. The evolutionary survival of the host is impaired, often very seriously, and we must enquire why this is so. Clearly, the crop pathosystem is unbalanced and the loss of balance must be due to an alteration in the host, in the environment, or in the parasite.

The host has clearly been altered; cultivars differ from wild plants and, in the course of domestication, have lost resistance.

The environment has also been altered; cultivation differs from a wild ecosystem. Host population densities, levels of nutrients and growth rates are unnaturally high; farming systems, cultivation techniques and other factors can contribute to increased damage from parasites. In so far as these factors contribute to high levels of damage, the crop pathosystem will require a higher level of horizontal resistance than does the natural pathosystem if the natural level of damage or disease is to be maintained in the crop pathosystem.

The parasite can also change, but only to a slight extent. For example, a high host population density may enable it to increase its horizontal pathogenicity at the expense of other survival values contributing to the reproductive rate or effectiveness of dispersal, but in general, the parasite retains all the optima of the natural pathosystem. Unlike the host, it is wild, not domesticated; it is not protected by man and can survive only in spite of man; to survive it must retain all those survival values which contribute to fitness. The very fact of its survival sets a limit to its horizontal pathogenicity or parasitism.

This is not to say that high levels of parasitic damage can never occur in the natural pathosystem. Evolution involves many thousands of seasonal epidemics which inevitably vary and exhibit an occasional extreme. The resilience of the pathosystem ensures that a recovery of systems balance and evolutionary survival is not impaired. Sporadically high levels of damage can also occur in accordance with the Rossetto hypothesis. The artificial level of damage is different in that it is continuously high, therefore it can only occur in an artificial system which is protected by man.

5.5.3 The Constant Sum of Survival Values

It is now necessary to state a general hypothesis which is that the sum of all variable survival values is a constant; that is, one survival value can be increased only at the expense of one or more other survival values. This hypothesis is applicable at all systems levels.

At the evolutionary level, we recognise that survival values must be variable in order to provide systems resilience. The climate is variable and the entire system must be able to respond to swings away from the norm and to recover from such swings. At the evolutionary level, all these variables collectively make up one survival value which is called fitness. As competition ensures that only the fittest survive, fitness must be at its maximum; any loss of fitness will lead to evolutionary extinction. In its turn, this means that every variable contributing to fitness must be at its optimum, as a variable which is either above or below the optimum detracts from fitness. It is common knowledge that unnecessary survival values tend to be lost due to negative selection pressure, which we can only explain by postulating that unnecessary survival values are not just neutral, but hindrances, hindering by their effect of decreasing other, necessary survival values.

A few examples will illustrate this point. Fungal and bacterial pathogens which are facultative parasites exhibit an inverse correlation between saprophytic ability and parasitic ability. It is common knowledge that cultures lose pathogenicity on nutrient agar; parasitic ability can be restored by serial passages through a host, but there is then a corresponding loss of saprophytic ability. The same phenomenon occurs with soil-borne pathogens and often necessitates crop rotation. Similarly, vertical and horizontal pathogenicities may be inversely correlated; it seems that one can only be increased at the expense of the other. The mechanism of strong vertical genes can only be explained on the basis of a reduction of other survival values. Finally, the process of domestication clearly indicates that survival values contributing to yield and quality can only be increased at the expense of others contributing to the fitness to compete in a natural ecosystem.

At the level of the individual organism, some survival values are mutually exclusive. With industrial melanism, for example, it is impossible for a moth to be both black and white; blackness can only be increased at the expense of whiteness, and *vice versa*.

At the molecular level of the pathosystem we are reduced to speculation which must be entirely theoretical. It is not impossible that the total genetic code is a constant and that an increase in one item of code is possible only at the expense of some other item of code. However, two qualifications are necessary.

Firstly, we must distinguish between genetic code, which is structure or spelling, and genetic information, which is behaviour or meaning. It is possible to increase the code without necessarily increasing the information. For example, we could send a telegram saying "The king is dead". If, soon afterwards, we sent a second telegram saying "The king is dead", we would have doubled the amount of code, or spelling, without any increase in information or meaning. Similarly, with the formation of an autotetraploid, the total genetic code is doubled without any increase in genetic information. In other words, when we postulate that the total genetic code is a constant, we really mean that the total genetic information is a constant and that one item of information can be increased at the expense of some other item of information.

Secondly, this discussion clearly refers to micro-evolution and historical time. Given geological time and macro-evolution, increases in genetic information are clearly essential because macro-evolution is impossible without them. Theoretically, an increase in genetic information can occur in two ways; it could be due to an increase in total code, provided that it is different code; that is, by the development of new chromosomal material by increases in either chromosome length or number. Alternatively, there could be an increase in the efficiency of the code. By analogy, a book might run to a second edition with an increase in its total information. This increase could be obtained by an additional number of pages or, alternatively, by an improvement in the quality of the writing. But in the natural evolutionary system, significant increases in genetic information require geological time, and any increases which might occur during historical time are likely to be both rare and of very minor significance.

5.5.4 *The Evolutionary Need for Susceptibility*

In this context, it is perhaps instructive to enquire why host susceptibility should occur at all. Every higher plant is immune to some millions of species of microorganisms and other potential parasites. That it should possess mere resistance to its actual parasites tends to be taken for granted because we take the whole phenomenon of parasitism for granted. But parasitism is the exception, with perhaps only one actual parasite species for every million potential parasite species. It is tempting to suggest that comprehensive immunity (i.e. no parasites whatever) is an unnecessary survival value, and that, by reducing a small amount of immunity to mere resistance, the host gains other survival values and, hence, an increase in its overall fitness.

5.5.5 Host Erosion of Horizontal Resistance

From this discussion, we can conclude that the main cause of cultivar susceptibility to pests and pathogens is due to a host erosion of horizontal resistance. We shall see later that a host erosion of resistance can be restored by breeding. Once the natural level of horizontal resistance is restored, the natural level of damage will also be restored except in so far as the crop environment and its effects on the parasite populations contribute to higher intensities of damage. However, if necessary, these other contributory factors can be compensated by a further domestication of horizontal resistance. This is the process of extending the series of horizontal pathodemes in the direction of increased resistance (Fig. 12).

5.5.6 Parasite Erosion of Horizontal Resistance

We must now consider the possibility of a parasite erosion of horizontal resistance. This would mean that, as we extend the series of horizontal pathodemes in the direction of increased resistance, nature would extend the series of horizontal pathotypes in the direction of increased parasitic ability (Fig. 12). If we are to postulate that this cannot happen, we must postulate first, that there is an effective maximum of horizontal parasitic ability and second, that all important pest and pathogens already possess that maximum.

The evidence that there clearly is an effective maximum to horizontal parasitic ability is irrefutable and is in three categories of descending order of importance. (1) Were there no limit to horizontal parasitic ability, we would have superparasites with respect to both host range and destructiveness; however, the world is still quite green. (2) The natural pathosystem can only be balanced or it would not have survived. This balance is the natural level of host damage which does not

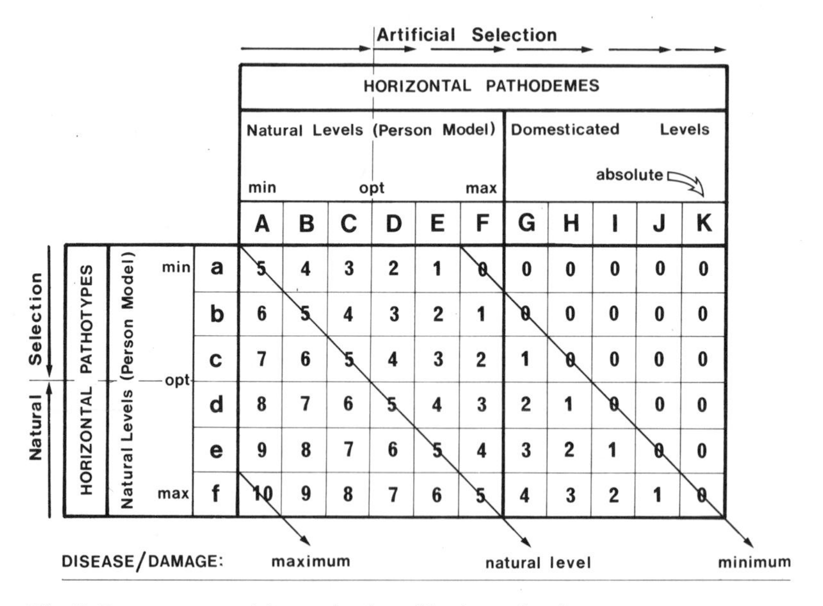

Fig. 12. Permanence and domestication of horizontal resistance

impair its evolutionary survival, and results from the interaction between the natural level of resistance and the natural level of parasitic ability. (3) The natural level of parasitic ability is the effective maximum. If we accept the hypothesis concerning the constant sum of survival values, the parasite can only gain parasitic ability at the expense of other survival values. Although the conditions of cultivation may permit such a change of balance, it can only be a small change and there is still a limit to its extent.

In general, we can argue that major crop parasites already possess the maximum horizontal parasitic ability, otherwise they would not be major parasites. In detail, there are two considerations. We must examine the extent to which a parasite can change its parasitic ability away from the optimum, and also the extent to which such changes have occurred in the past or may occur in the future.

The general rule is that facultative parasites can change a lot and that obligate parasites can change very little. Parasitic ability and saprophytic ability are inversely correlated and a facultative parasite usually exhibits reversible changes from the one extreme of being a near-obligate saprophyte to the other extreme of being a near-obligate parasite, its parasitic ability being maximal when its saprophytic ability is minimal. An obligate parasite (e.g. *P. infestans*), as is well known, can change very little and it is also obvious that a change in either direction away from the natural optimum will gravely impair evolutionary survival in an obligate parasite.

We could postulate that the artificially high levels of damage in the crop pathosystem are due, at least in part, to past increases in parasitic ability. We would then be forced to conclude that such parasites now have a parasitic ability above the natural optimum and, consequently, now have a parasitic ability which is at or near the maximum. Conversely, if we are to postulate future increases in parasitic ability, leading to major parasite erosion of horizontal resistance, we must draw one of two conclusions. Either the parasitic ability is now below the natural optimum or major increases above the optimum are possible.

What does all this mean in practice? It means that, if we are breeding for horizontal resistance to an obligate parasite, the danger of a parasite erosion of horizontal resistance is slight, and if it occurs, its effect will be small. If we are breeding for horizontal resistance to a facultative parasite, we must ensure that our screening for resistance is conducted with a parasite population exhibiting minimum saprophytic ability and maximum parasitic ability. In other words, given sound pathosystem management, there seems to be little danger of a loss in the effectiveness of horizontal resistance due to increases in parasitic ability.

Chapter 6 Horizontal Pathosystem Management

This Chapter concerns horizontal pathosystem management: how to demonstrate horizontal resistance, measure it, assess its usefulness in agriculture and, above all, breed for it.

6.1 The Demonstration of Horizontal Resistance

Throughout this Section it must be remembered that horizontal resistance can only be demonstrated in the esodemic. This means that vertical resistance, if it is present, must have broken down.

6.1.1 Constant Ranking

Demonstration of a constant ranking of pathodemes, according to resistance, regardless of which pathotype they are tested against; and a constant ranking of pathotypes, according to pathogenicity, regardless of which pathodeme they are tested against, is possibly the best evidence for the horizontal nature of a pathosystem. The constant ranking means that the resistance and pathogenicity are independent of each other. There is no gene-for-gene relationship.

There are three categories of evidence for constant ranking: experimental, historical and geographical.

An experimental demonstration of constant ranking requires a series of different pathodemes and a series of different pathotypes, and, in each series, the differences must be statistically significant. This demonstration is not always an easy one, and there is a spectrum of difficulty depending on the pathosystem in question. At its easiest, the demonstration involves a host, such as sugarcane, in which vertical resistance does not occur, and a facultative parasite which has a wide range of differences in horizontal parasitic ability. The difficulties increase with vertical resistance in the host, which must then be eliminated, at least for experimental purposes, either by using pathodemes with no vertical genes, as in potato blight, or by using pathotypes which match the vertical resistance in question. Incomplete vertical resistance adds further problems. If the parasite is an obligate parasite, differences in horizontal parasitic ability are small and it is difficult to obtain a series of different horizontal pathotypes. However, this problem can sometimes be solved by using other *formae speciales* of the same parasite which may have a low horizontal parasitic ability on the test host, and parasite hybridisation may also be possible. Lastly, there may be a differential interaction which is not due to vertical resistance. It is clear that, in some breeding programmes, an experimental demonstration of constant ranking is too difficult to justify. Ironically, these are likely to involve crops in which vertical resistance has been the most mis-used and in which horizontal resistance is the most urgently required.

Historical evidence of constant ranking comes from an examination of old cultivars, if any are available. Ideally, these cultivars should date from before any vertical resistance breeding of the crop was started. Van der Plank (1971) demonstrated that the ranking of old Dutch potato cultivars for resistance to *P. infestans* had not changed during thirty years. Similar demonstrations are possible in many crops and make useful student exercises. However, a historical demonstration of constant ranking is subject to all the difficulties listed above.

Geographical evidence for constant ranking is obtained by field testing the series of different pathodemes in various parts of the world. The overall levels of disease may vary widely but this does not matter provided that the pathodeme ranking is constant, a demonstration valuable if it can be made. Often, however, it cannot be made due to the various problems listed in the discussion on experimental demonstration. Of these, a differential interaction between horizontal resistance mechanisms and environmental factors is probably the most important. It apparently occurs, for example, in sugarcane smut, bacterial blight of cotton, Phomopsis disease of tea and ratoon stunting disease of sugarcane (7.3).

A final and obvious comment on the demonstration of constant ranking is that inability to demonstrate it is not proof of a lack of horizontal resistance.

6.1.2 *Genetical Evidence*

It seems that polygenically inherited resistance is always horizontal. If, therefore, a breeder is confident that the resistance in question is inherited polygenically, he can be confident that it is horizontal resistance, as the possibility of vertical resistance being inherited polygenically is so remote that it can be disregarded. This is the most useful evidence for the horizontal nature of resistance in that it will be the most frequently used. Occasionally, horizontal resistance may be inherited oligogenically but this phenomenon is rare and each occurrence will have to be treated on its own merits.

6.1.3 *Mechanisms*

If a resistance mechanism is manifestly beyond the capacity for micro-evolutionary change of a parasite, it confers horizontal resistance; a situation occurring occasionally and most commonly with oligogenic horizontal resistance. More commonly, horizontal resistance is conferred by many different mechanisms, most of which are complex and can vary in degree. Horizontal resistance is demonstrated, therefore, by mechanisms which collectively reduce the rates of host infection and colonisation, and parasite reproduction.

6.1.4 *Epidemiological Evidence*

Horizontal resistance slows down the epidemic (van der Plank, 1963); more specifically, it slows down the esodemic. If it can be conclusively demonstrated that vertical resistance is present and that it has broken down, any remaining resistance is horizontal.

Vertical resistance can confer apparent horizontal resistance in various ways, which can be distinguished epidemiologically if the role of the exodemic can be recognised and analysed.

6.1.5 The Nature of the Resistance

Horizontal resistance is usually (but not necessarily) quantitative resistance and the quantitative nature of the resistance is useful evidence in itself. Unfortunately, although vertical resistance is usually qualitative resistance, it can also be quantitative when it confers incomplete protection against non-matching vertical pathotypes. However, quantitative horizontal resistance often increases with increasing age of the host individual (adult plant resistance) while quantitative vertical resistance tends to be constant at all ages (juvenile or seedling resistance). The two kinds of quantitative resistance can also be distinguished by their inheritance; quantitative horizontal resistance is always polygenic resistance and quantitative vertical resistance is always oligogenic.

6.1.6 The Nature of the Host

As we have seen, vertical resistance is more common in host species with a high spatial and sequential discontinuity of host tissue, which means that it occurs most frequently in temperate annuals and least frequently in tropical perennials and it is likely to be totally absent from tropical, evergreen perennials. In some hosts, a breeder can be confident that vertical resistance does not exist and that, consequently, he can only be working with horizontal resistance.

6.1.7 The Nature of the Parasite

It has been argued that vertical resistance against bacterial and virus pathogens and insect parasites is rare. This is a rule, not a scientific law and its generality will doubtless be proved by exceptions in due course. Meanwhile, breeding for resistance to bacteria, viruses and insects is likely to involve horizontal resistance only.

6.1.8 The Breeding Technique

If we have accumulated horizontal resistance by breeding under conditions such that vertical resistance, or its effects, were eliminated, we can be confident of the horizontal nature of the resistance. As our techniques and confidence improve, this category of evidence is likely to make any other demonstration unnecessary.

6.1.9 Negative Evidence

It is extremely unlikely that any host has more than one vertical resistance mechanism to any one parasite, but again, this is a rule and exceptions may yet be found. However, if vertical resistance is known, its absence is negative but clear evidence for the horizontal nature of the resistance, which means that the various tests for the vertical nature of resistance should all be negative.

6.2 The Measurement of Horizontal Resistance

We shall discover that it is often unnecessary to measure horizontal resistance, as its accumulation can often be assumed from the breeding technique (6.4) and, when there is enough horizontal resistance to control the disease, its measurement is somewhat academic. Nevertheless, it may be necessary to measure levels of horizontal resistance for research purposes, pathosystem analysis and the demonstration of constant ranking.

Horizontal resistance can only be measured in terms of the interaction between the parasite and the host; that is, the amount of damage or disease. The amount of damage is a measure of resistance if the pathogenicity is known, and a measure of parasitic ability if the resistance is known.

It must be repeated that horizontal resistance can only be measured in the esodemic; that is, after vertical resistance, if it occurs, has broken down. In other words, in vertically resistant crops, horizontal resistance can only be measured with a matching vertical pathotype.

6.2.1 *Relative Measurements*

The ranking of horizontal pathodemes, according to resistance, is itself a measurement of that resistance. One cultivar is more resistant than another, but perhaps less resistant than a third cultivar. These relative measurements are normally obtained with field trials in which all variables are controlled statistically. The resistance is measured in terms of the amounts of the disease or damage at the time of harvest. The most important source of error is interplot interference (van der Plank, 1963; James et al., 1973) in which the level of damage in one plot is affected by neighbouring plots. Attempts to measure a number of different parameters, such as rates of host infection and colonisation, and parasite reproduction are likely to become excessively complicated. The best field techniques for the assessment of damage are to be found in the FAO Manual of crop loss assessment methods (Chiarappa, 1971) and the Canadian manual on the same subject (James, 1971, 1973).

6.2.2 *Absolute Measurements*

Absolute measurements of horizontal resistance require laboratory conditions in which all other variables are constant. Such assessments can be very precise but they are also complicated. There may also be problems associated with relating absolute measurements to field performance in that a field environment may differ markedly from the standard laboratory conditions.

6.2.3 *Scales of Measurement*

Few scales of measurement of horizontal resistance have been defined because little work has been attempted with this form of resistance. In potato blight, the scale is usually taken as 0–4 where 0 is no disease and 4 is complete destruction. Strictly, this is a scale of susceptibility rather than resistance. It measures the damage done by the pathogen.

In theory, there are six points which can be defined on a horizontal resistance scale of measurement.

1. Is absolute susceptibility: a complete absence of horizontal resistance such that the parasite can grow through the host tissues without any hindrance whatever. This point does not exist in practice but it might, perhaps, be inferred by comparison of the parasite growth at the absolute minimum horizontal resistance (see below) and parasite growth in the ideal culture medium.

2. Is the absolute minimum horizontal resistance. This is the minimum of variable horizontal resistance; what resistance remains is non-variable and is due to cell walls, etc. In theory, the absolute minimum can be determined experimentally by imitating the African maize situation (p. 96) in reverse. That is, there should be strong selection pressure for susceptibility during some 10–15 host generations of random polycross. In practice this may prove difficult and, at the very least, would require a technique for eliminating the parasite to ensure individual host survival once the susceptibility had been determined.

3. Is the unselected level of horizontal resistance. This can also be determined experimentally be permitting 10–15 generations of random polycross in the absence of the parasite.

4. Can be defined with accuracy but is difficult to determine experimentally. It is the natural level of damage and is that level at which the evolutionary survival of the host in a natural pathosystem is not impaired.

5. Is the maximum variable horizontal resistance. This can be determined experimentally and involves selection pressure for horizontal resistance during some 10–15 generations of random polycross in the absence of all other selection pressures.

6. Is absolute horizontal resistance. This is equivalent to immunity but, is an apparent immunity because it can be host eroded. It may be only a theoretical possibility with some parasites.

6.2.4 Comprehensive Resistance

Normally, horizontal resistance is measured with respect to a particular parasite species. If a crop has, say, twenty parasites, then twenty assessments of horizontal resistance would be necessary. Later, we shall discuss the development of comprehensive horizontal resistance, which involves a higher level of the pathosystem (9.1) in which resistance to all locally important parasites is treated as a single survival value and a single selection criterion. Measurement of horizontal resistance then involves only the one parameter of parasite damage, regardless of what parasites may be causing that damage. It may be possible to operate at an even higher system level at which parasite damage is regarded as one of many variables contributing to yield and/or quality which are then treated as single survival values and single selection criteria. This holistic approach will be anathema to some scientists but it does emphasise that, given the correct breeding technique, it is unnecessary to demonstrate horizontal resistance or to measure it, except at the highest systems levels. In purely practical terms, this can lead to great economies in research resources and corresponding increases in the amount of screening per unit of skilled manpower.

6.3 The Value of Horizontal Resistance in Agriculture

When considering the usefulness of horizontal resistance in agriculture, it is convenient to discuss its limitations first and then its advantages.

6.3.1 The Limitations of Horizontal Resistance

On theoretical grounds, Robinson (1973a) suggested that the usefulness of horizontal resistance was inversely proportional to the commerical value of the crop. We can visualise a spectrum from the most primitive to the most advanced forms of agriculture. At the one extreme is the exploitation of wild plant populations, including food gathering by primitive people, timber collecting from wild forests, ranching of natural grasslands and, in the animal world, sea fishing. Next in the spectrum is subsistence agriculture, followed by extensive commercial farming, intensive farming and horticulture, and ending at the extreme of factory farming in automated glasshouses. Through the spectrum, the amount of artifical selection in the host increases and also the yield and quality of the product, the value of the product, and, hence, the permissible expenditure on its production. But at the same time, both the level and the usefulness of horizontal resistance tend to decrease through the spectrum, horizontal resistance being normally at its highest in wild plants and at its lowest in intensely selected, specialised cultivars. At the one extreme, cheapness is essential and horizontal resistance is imperative, at the other, quality is essential and the cost of an artificial control of parasites may be preferable to the difficulties of developing adequate horizontal resistance.

Horizontal resistance may have other limitations, in particular the natural level of horizontal resistance may be inadequate to control a parasite in the artificial, crop pathosystem. There are six possible reasons for this.

1. In the natural pathosystem, the crop progenitors may have possessed a marked, apparent horizontal resistance due to a mixture of vertical resistances. Their true horizontal resistance would be correspondingly reduced and the natural level may be inadequate to control the parasite under conditions of agricultural crop uniformity.

2. The natural pathosystem may have involved low population densities of the host. The natural level of horizontal resistance would then be inadequate under conditions of agricultural crowding.

3. There may have been a separate evolution of the host and the parasite (8.2). It may then be difficult to find adequate horizontal resistance.

4. The more extreme environments of agriculture may be so far removed from the climatic optima of the natural pathosystem that the natural level of horizontal resistance is inadequate.

5. The Rossetto hypothesis may be a factor in minimising the natural level of horizontal resistance. The natural restraints on parasite reproduction may be missing in the crop pathosystem.

6. Artificially high levels of horizontal resistance may be difficult to preserve in some crops, such as subsistence maize.

Against all these arguments we can pitch the possibility of the domestication of horizontal resistance (9.2). We must recognise, however, that this will be more difficult in some crops than others.

Finally, we must recognise a special artificiality in agriculture. Towards the end of a season and, hence, the end of the epidemic, much host tissue becomes expendable, including all parts of an annual plant except the seed and its protective coverings, all the leaves of a deciduous tree and the older leaves of an evergreen tree, and edible fruit tissues which assist seed dispersal by animals. Such expendable tissues are unlikely to have high levels of resistance in a natural pathosystem because that resistance would have no survival value. It is common knowledge that senescent host tissue not only loses resistance to parasites but that it is increasingly invaded by saprophytes. Occasionally, in agriculture, it is the naturally expendable tissues which are the harvestable product, particularly so in the more luscious fruit tissues and in these it may prove very difficult to develop agriculturally adequate resistance.

6.3.2 The Advantages of Horizontal Resistance

At this stage, we need only note the three basic advantages of horizontal resistance in agriculture:

1. Horizontal resistance can be domesticated, just as yield and quality have already been domesticated, which means that the artificial levels of horizontal resistance can be higher than the natural levels (see 9.2).

2. Horizontal resistance permits crop uniformity which we must recognise as an essential prerequisite of high productivity (see 9.2).

3. Breeding for horizontal resistance is cumulative; a good cultivar is replaced only with a better cultivar. This is in contrast to the traditional repetitive breeding for vertical resistance in which "it takes all the running you can do to stay in the same place." Cumulative crop improvement is the subject of Chapter 9 in which it will be seen that Hegelian changes in improvement become possible.

6.4 Maize in Africa

It is now necessary to discuss maize in Africa with particular reference to *P. polysora*. Much of the discussion concerns pathosystem analysis, but this cannot be conveniently separated from the aspects of pathosystem management which are more important and which belong to the present Chapter.

Maize *(Zea mays)* was first imported to Africa some four centuries ago. It was re-imported many times, by many routes and in many forms. It was imported as seed, as grain for consumption, either hulled or on the cob, and as straw for packing material. For most of this period, the infectious nature of plant diseases was unknown and phytosanitary precautions were never contemplated, so that virtually all the maize parasites were imported also, the only restriction on parasite import being the time factor. Sea transport was slow, particularly with sailing ships, and short-lived parasites could not survive the journey. *P. polysora* was apparently such a short-lived parasite, and was not imported alive until after the development of trans-Atlantic air transport.

Although there is considerable commercial cultivation of maize in Africa, the bulk of the cultivation is in the form of subsistence food crops grown by peasant farmers. Being an out-breeding annual, maize is highly responsive to selection pressures, particularly under peasant farming conditions in which each farmer uses some of his own crop for seed. Local landraces develop quickly and these are well adapted to the wide variety of local environments which differ in many climatic and geographical variables throughout Africa, affecting the parasite as well as the host and, in areas where a parasite was favoured, the host responds to the increased selection pressure by accumulating higher levels of resistance. The maizes of Africa are thus in a good state of pathosystem balance and approach a natural pathosystem.

6.4.1 P. polysora

P. polysora was introduced to West Africa in the 1940's, apparently on green cobs imported by air (Cammack, 1959). It reached East Africa in 1952 (Nattrass, 1952). In any one area, the effects of *P. polysora* had a characteristic pattern. The initial amount of destruction declined steadily to a base level where it now remains. The initial damage was often so great that there were justifiable fears of a major famine comparable to the Irish potato famine of the 1840's caused by *P. infestans*. These fears proved unfounded for two reasons. (1) *P. polysora* has clear climatic limits. At the equator in Africa it lacks epidemiological competence above 4000 ft and its altitude limits decrease with increasing degrees of latitude. (2) As we have seen, the levels of damage decline steadily to a base level which is not destructive. Nevertheless, even if the fears of famine proved unfounded, they were very real at the time and resistance breeding programmes were hurriedly initiated in a number of countries.

6.4.2 The Work on Vertical Resistance

The West African breeding work has been described by Stanton and Cammack (1953) and the East African programme has been described by Storey *et al.* (1958).

During the 1950's the fashion for the gene-for-gene relationship was reaching its height. It was inevitable that the maize-breeding programmes should involve oligogenically inherited hypersensitivity, and even the possibility of an alternative was apparently not considered. Vertical genes to *P. polysora* are rare and, as Storey *et al.* (1958) commented, none were found in the African maizes. (This supports the contention that vertical resistance is not necessarily an important survival value in a heterogeneous population and that it need not necessarily evolve. Hooker and le Roux (1957) obtained similar results with *Puccinia sorghi*). However, two vertical genes, Rpp1 and Rpp2 were discovered in maize lines imported from the Americas. Various complications, such as incomplete vertical resistance, need not concern us here, but what is important is that the vertical resistance broke down so quickly that it was valueless. By the time this was appreciated, *P. polysora* was no longer a serious disease. As we now realise, the host had accumulated horizontal resistance and vertical resistance was not only valueless, it was unnecessary.

6.4.3 *Imitation in Other Crops*

Van der Plank (1968) was the first to comment that the decline in the severity of *P. polysora* in Africa was due to an accumulation of horizontal resistance and that it showed "how quickly cross-pollination and selection pressure can accumulate resistance in a hetereogeneous population provided that it has an adequately broad genetic base". Robinson (1973a) commented that an analysis of the factors which governed that accumulation would provide guide-lines for imitating the maize situation in other crops. It would, in fact, reveal the factors which are important in any breeding for horizontal resistance. The remainder of this section concerns those factors.

6.4.4 *Parasite Erosion of Horizontal Resistance*

Initially, the *P. polysora* epidemics were very destructive, but the pathogen had come from the Americas where it was not destructive. The improbability of pathogen erosion of horizontal resistance, particularly with an obligate parasite, becomes apparent. If the initial disease severity was due to a gain in horizontal pathogenicity, that gain could only have occurred during a short air journey across the Atlantic. If we accept this remote possibility for the purposes of discussion, we must then explain the subsequent decline in disease severity, which on this basis, could only be due to a loss in horizontal pathogenicity; let us accept this remote possibility also. We must then explain why the pathogen gained horizontal pathogenicity every time it reached a new area within Africa, only to lose it again a few years later. We must also explain why, when high altitude maizes are taken to *P. polysora* areas, the pathogen gains horizontal pathogenicity during the one short journey from the local maize to the high altitude maize. We must conclude that the initial severity of *P. polysora* was not due to changes in horizontal pathogenicity and that, in general, parasite erosion of horizontal resistance is not an important factor with obligate parasites.

6.4.5 *Environment Erosion of Horizontal Resistance*

The same argument applies to a possible environment erosion of horizontal resistance. This might be due to differences of natural environmental factors, such as soil and climate—the second corner of the disease triangle. It might also be due to artificial environmental factors, such as cultivation methods—the third corner of the disease square. No doubt, these factors do differ from those in the Americas where *P. polysora* is not destructive, but we must then explain the subsequent decline in disease severity. Neither the climate nor the soils of Africa nor the deeply ingrained peasant cultivation traditions changed. We must conclude that an environment erosion of horizontal resistance was not an important factor.

6.4.6 *Host Erosion of Horizontal Resistance*

The only valid explanation of the decline in severity of *P. polysora* involves the host—the sole remaining corner of the disease square. From the extreme improbability of pathogen erosion and environment erosion, we come to the extreme probability of host erosion of horizontal resistance.

Host erosion of horizontal resistance is due to genetical changes in the host population, and is the result of negative selection pressure; that is, it occurs when horizontal resistance is an unnecessary survival value. It occurs, typically, in the absence of the parasite, as clearly happened with the African maizes in the absence of *P. polysora*. It may also occur during domestication when other survival values are emphasised, or during breeding for yield and quality, when artificial disease control measures are emphasised, or during the breeding for vertical resistance, when it is known as the vertifolia effect.

The converse of host erosion is the host accumulation of horizontal resistance. This is the result of positive selection pressure; that is, it occurs when horizontal resistance is a necessary survival value. It occurs in the presence of the parasite when other survival values, artificial control measures and vertical resistance are not interfering.

It is clear that the initial severity of *P. polysora* was due to a host erosion of horizontal resistance which occurred in the absence of the parasite. The subsequent decline in severity was due to a host accumulation of horizontal resistance which occurred in the presence of the parasite under peasant farming conditions in which artificial disease control measures and vertical resistance were absent. These host population changes were due to reversible alterations in gene frequencies.

We can now recognise that host erosion of horizontal resistance is so common in domesticated plants as to be almost universal. It is the main reason that cultivars are generally more susceptible to parasites than wild plants and that the crop pathosystem is generally more destructive than the natural pathosystem.

The most famous example of host erosion was in potatoes, which were introduced to Europe more than two centuries before *P. infestans*. During this period, potatoes were domesticated, more or less unconciously, by man. One of the main effects of the domestication was the increase in day-length and they did not become a major crop until this was achieved. A second effect was the host erosion of horizontal resistance to *P. infestans*. When this parasite appeared, its destructiveness virtually initiated the study of plant diseases. Two misconceptions of the period are relevant.

The first was the complete ignorance of virus diseases, although it was well known that potato stocks would decline in vigour. This was attributed to some mystical effect of vegetative propagation which was considered an unnatural and harmful form of reproduction. The decline in vigour was overcome by repeated renewal of stocks, that is, by producing new cultivars from true seed. There was much sexual reproduction, even if the overall genetic base was narrow. Then blight appeared and it was observed that its severity declined after a few years. The second misconception attributed this to a mystical loss of "virulence" in the parasite, an interpretation widely accepted but never explained. We now appreciate that the decline in severity was due to genetical changes in the host population. The most susceptible clones were totally destroyed; less susceptible ones were abandoned as cultivars: only the most resistant clones continued to be cultivated. Later, new clones were produced and blight resistance was a primary selection criterion. Gradually, the level of horizontal resistance increased until, some 40 years after blight appeared, Millardet discovered Bordeaux mixture. The absolute necessity for resistance then vanished and we must recognise that Millardet's discovery postponed the further accumulation of horizontal resistance to blight for nearly a century.

Van der Plank (1975) has commented that an excess of horizontal resistance is probably harmful and that this is why any excess tends to be lost due to negative selection pressure. This is in accordance with the hypothesis of a constant sum of variable survival values. Unnecessary survival values are hindrances because they detract from other, necessary survival values. However, van der Plank was referring to pathosystems in which the control was autonomous, when he quotes the loss of resistance to *P. polysora* in African maizes and the loss of resistance to *P. infestans* in European potatoes. When the pathosystem control is deterministic, however, horizontal resistance can be domesticated, which will no doubt lead to a loss of fitness to compete in a natural ecosystem. But this loss will be no more important than comparable losses due to the domestication of yield or quality. Cultivars will not survive in a natural ecosystem anyway; nor is it necessary for them to do so; but it is very necessary that cultivars should be domesticated in all their agricultural qualities, including their horizontal resistance.

A final comment concerns the apparent erosion of horizontal resistance. Occasionally, there may be a gradual change in the farming system. Rotation may be reduced, or the burning of crop residues may be abandoned. There will then be an increase in the levels of disease. This is what we mean by environment erosion of horizontal resistance. It would be all too easy to postulate, falsely, that the increase in disease was due to genetical changes in the parasite population (parasite erosion) or in the host population (host erosion).

6.4.7 The Unselected Level of Horizontal Resistance

The unselected level of horizontal resistance has already been discussed: it is the low level of variable horizontal resistance which is reached in the absence of selection pressure for it; but considerable horizontal resistance must still remain. Some resistance mechanisms serve a dual function, including such physical, passive mechanisms as the cuticle and cell walls. It is also likely that some unspecialised resistance mechanisms operate against many parasites and selection pressure for them would continue in the absence of only one parasite.

The unselected level of horizontal resistance is a natural balance for the state of the system–that is, the system in the absence of the parasite. It also represents the natural limit to host erosion of horizontal resistance. There is no apparent reason why it should not be determined experimentally. We must also presume that the absolute minimum could also be determined experimentally with an artificial, positive selection pressure for susceptiblity. Such data would provide useful, fixed points on a measurement scale.

Possibly the most important aspect of the unselected level of horizontal resistance is that it represents a very high agricultural susceptibility. This means that, susceptible though they may be, most cultivars possess at least the unselected level of horizontal resistance and possibly more.

6.4.8 *Sources of Horizontal Resistance*

The African maizes possessed the unselected level of horizontal resistance. There was then a host accumulation of horizontal resistance which occurred entirely within those maizes; effectively, there was no introduction of foreign parent material to their gene pool. It must be stressed that the work on vertical resistance is prominent in the scientific literature, but that the foreign maizes which it introduced never became prominent in the many millions of acres of peasant maize. Indeed, most of this imported material never went beyond the experiment stations.

One of the characteristic features of traditional breeding for pest and disease resistance is that the first step has always been to find a "good source" of resistance, without which, it was argued, resistance breeding is impossible. The good source of resistance was then transferred to cultivars by back-crossing; and this was so much easier if the resistance was inherited oligogenically. In the African *P. polysora* breeding, which was traditional, the workers looked for (vertical) resistance genes in the African maizes but none were found; they had to be imported from the Americas.

It is now clear that we can accumulate horizontal resistance by working exclusively with high-yielding, high-quality but susceptible cultivars and that it is not necessary to search for an external source of resistance. At the very least, it is reasonable to attempt this approach and to search beyond existing cultivars only if it proves impossible to accumulate adequate horizontal resistance within them.

6.4.9 *Freedom from Masking Factors*

Any factor which masks horizontal resistance will reduce selection pressure for it, masking meaning any factor which reduces the esodemic. The African maizes were notably free from factors masking the *P. polysora* esodemic. The most obvious masking factor is a foliar fungicide. There can be no selection pressure for resistance if every spore, however dense the spore load, is killed before infection begins. It follows that all screening for horizontal resistance should be conducted in the absence of artificial disease control measures, including the less obvious ones such as rotation and the burning of crop residues.

In some crops, the most insidious of these masking factors is vertical resistance which, in the African *P. polysora* epidemic was either absent or insignificant. As we have seen, horizontal resistance only functions in the esodemic, after vertical resistance has broken down. Furthermore, if there is a mixture of different vertical resistances in the screening population, there will be an apparent horizontal resistance. This will reduce selection pressure for true horizontal resistance, and its effects will be lost when host population uniformity occurs in commercial cultivation. It follows that in, say, wheat or rice, either the vertical resistance itself or its effects must be eliminated during the screening process.

Vertical resistance, or its effects, can be eliminated in four ways. Vertical resistance itself can be eliminated genetically by ensuring that all the original parents of the screening population possess no vertical genes. This is quite easy in a crop such as potatoes where there are many cultivars which lack vertical genes, but it is likely to prove impossible in crops such as wheat, rice, barley or oats.

Secondly, vertical resistance can be eliminated by a negative screening of all individuals showing evidence of it. Again, this is easy with a crop such as potatoes where *P. infestans* produces a hypersensitive flecking on vertically resistant leaves, and it is also possible to screen for "slight disease" rather than "no disease" when vertical resistance provides a complete protection against non-matching vertical pathotypes. Another possibility is to conduct a seedling screening and to discard all individuals with juvenile resistance. These techniques are all labour-intensive, however, and can be very wasteful of host material in the early stages of the screening process.

The effects of vertical resistance can be eliminated epidemiologically by choosing a vertical pathotype first and then using as many cultivars as possible which have full vertical susceptibility to it. Each screening epidemic is then induced artificially using that vertical pathotype as the initial inoculum. This is probably the best approach for a crop such as wheat in which there are many vertical genomes and the whole picture is confused by incomplete vertical resistance. However, the possibilities of strong vertical gene effects and a reduced horizontal pathogenicity in a complex vertical pathotype should be borne in mind. It is probably safer to risk too narrow a gene base, than to encounter these complicated, epidemiological side-effects.

Finally, the effects of vertical resistance can be eliminated by the saturation approach in which a range of different vertical pathotypes are used in a mixture of different vertical pathodemes. This is essentially what happens with *P. infestans* in the Toluca Valley of Mexico, where functional oospores produce a wide variety of vertical pathotypes. This area has proved valuable as a screening location and a testing ground for potato breeders throughout the world. Because the vertical resistance of any potato cultivar breaks down quickly, the unknown horizontal resistance of the cultivar can be quickly assessed. However it is a very different matter trying to synthesise such an epidemic with other diseases in experimental fields. The main difficulty is one of uncertainty. Suppose the disease was wheat stem rust, for example; however wide the range of vertical pathotypes, there is always a possibility that certain host individuals will escape allo-infection entirely or be allo-infected very late in the epidemic. There will then be considerable doubt concerning the nature of the resistance in the final selections.

6.4.10 *Cross-Fertilisation*

Maize is an outbreeder and, until recently, it was difficult to achieve similar levels of cross-fertilisation with inbreeding crops such as wheat or rice. If we are to imitate the African maizes in such crops, we must achieve similar levels of cross-fertilisation. One of the most valuable new tools available to the pathosystem manager is the class of chemicals known as male gametocides. They can be used to make an inbreeder male-sterile and it can then be easily cross-pollinated. A male gametocide has four advantages over either genetic or cytoplasmic male sterility. It is easier to use; it can be applied to any individual or population; its effects are not inherited; and there are no undesirable, linked characters.

Male gametocides are now being investigated mainly with a view to facilitating hybrid seed production. One of the more promising appears to be "Ethrel" (2-

chloro-ethyl-phosphonic acid), (Rowell and Miller, 1971) but rapid development can be expected. Hybrid seed production requires a very high rate of male sterility and, so far, male gametocides are inadequate. The criteria in breeding for horizontal resistance are, however, much less strict. Lower rates of cross-fertilisation would be disadvantageous only in that they might necessitate a greater number of generations in the screening programme or larger host populations and the use of marker genes.

A theoretical possibility is that some individuals in the mixed populations may possess a greater tolerance than others to a male gametocide. There will then be strong selection pressure for such individuals and the effectiveness of the male gametocide will decline. The problem could be solved by using a sequence of different male-gametocides, or by using increasingly higher dosages. The use of marker genes (6.5) would also overcome this difficulty.

It must be commented also that genetic and cytoplasmic male sterility can also be employed and, in some crops, may be preferable to male gametocides.

6.4.11 Genetic Heterogeneity

As we have seen, maize was introduced to Africa many times and by many routes. It is thus very different from the world distribution of arabica coffee (6.7) and a wide genetic base is present in the overall African maize population, which is of interest in two ways.

In spite of the wide genetic base, all the maizes were highly susceptible to *P. polysora*. This perhaps emphasises that the phenomenon of host erosion of horizontal resistance is neither random nor rare, but an invariable and inevitable event in the absence of the parasite.

We must enquire how wide a genetic base is necessary in other crops if we are to accumulate horizontal resistance effectively. It seems that the genetic base can be far narrower than is generally realised. Most of the African maizes consisted of subsistence cultivars sown from seed of each farmer's own crop. This seed was consequently the result of many generations of random polycross and it is doubtful if any two maize plants were identical in any one farmer's crop. But each of these crops was a landrace in which all individuals possessed many characters in common. As such, it had a relatively narrow gene base, and the accumulation of horizontal resistance occurred within each farmer's crop. Undoubtedly, some cross-pollination occurred between crops and, possibly, some exchange of seed occurred between farmers, but the very wide genetic base of all the maizes of Africa is relevant only in that it demonstrates that the host accumulation of horizontal resistance in the presence of the parasite is as common and inevitable as is host erosion in its absence. There were perhaps a million peasant maize crops exposed to *P. polysora* in Africa; horizontal resistance accumulated in all of them. It also accumulated more or less independently in each of them. The accumulation of horizontal resistance was not only universal, it occurred within a relatively narrow gene base.

6.4.12 The Intensity of the Epidemic

When *P. polysora* reached Africa, the local maizes were highly susceptible. The most susceptible individuals were killed; less susceptible individuals were prevented from reproducing; still less susceptible individuals suffered a reduced reproduction. There was thus strong selection pressure for resistance. It should be noted that the selection pressure was exerted solely by the pathogen under conditions close to those of a natural pathosystem. After a succession of epidemics, resistance was restored to the natural level for the state of those pathosystems. Four points arise:

When imitating this situation in other crops, the epidemic (or infestation) should be as natural as possible, provided that the Rossetto hypothesis is not a factor. Artificiality can be deleterious principally because so many horizontal resistance mechanisms are involved and we know so little about them. To quote an obvious example, too early an epidemic might eliminate useful adult plant resistance.

The epidemic must be severe if it is to exert adequate selection pressure, but it must also be regularly severe, as a failure of the epidemic will waste a season and may lead to a host erosion of horizontal resistance. To this extent, the natural epidemic should be artificially assisted with simple cultural techniques. Depending on the type of parasite, these might include the avoidance of rotation or burning of crop residues; provision of an initial inoculum; an increased intensity of epidemic due to surrounding susceptible cultivars; an improved micro-climate with irrigation; and so on.

If vertical resistance is to be eliminated epidemiologically the epidemic must be artificial to the extent of providing an initial inoculum of the matching vertical pathotype.

As horizontal resistance accumulates, the intensity of the epidemic declines. The rate of accumulation of horizontal resistance then also declines, which is relevant in two ways. If the intensity of the epidemic is artificially maintained, fewer host generations will be necessary and the breeding programme will be completed sooner. Even more important, the decline in the intensity of the epidemic is related to the decline in the accumulation of horizontal resistance. A balance is reached which is the natural level of horizontal resistance for that pathosystem (i.e. it is natural in that there is no further selection pressure for resistance, in spite of those cultivation factors, such as host crowding, which tend to increase the level of damage). In this sense, the African maizes have reached the natural level of resistance and the disease is no longer damaging. However, if we wish, we can raise the level of horizontal resistance even higher by artificial screening in the course of artificially intensified epidemics. This is the domestication of horizontal resistance.

6.4.13 Epidemiological Competence

The fact that *P. polysora* exhibited strict ecological limits is relevant to pathosystem management in two ways.

First, we have already recognised the phenonemon of epidemiological competence, first defined by Crosse (1968). *P. polysora* is unable to cause an epidemic above 4000 ft of altitude at the equator. The maizes of these high altitude areas are thus suceptible to *P. polysora* but they are not vulnerable (8.1) to it. The limits of epidemiological competence are beyond the parasite's capacity for change, just as is horizontal resistance; there is a limit to the micro-evolutionary capacity for change of any species. No one seriously expects *P. polysora* to invade high altitude tropical areas, any more than he expects it to invade temperate latitudes. There is consequently no need for high altitude or high latitude maizes to possess horizontal resistance to *P. polysora*, which is why comprehensive horizontal resistance is defined as resistance to all *locally* important parasites.

Secondly, because high altitude and high latitude maizes lack horizontal resistance to *P. polysora*, they are valueless in areas where this parasite is epidemiologically competent. Thus development of comprehensive horizontal resistance necessitates many breeding programmes each suited to a particular pathosystem, and a particular distribution of parasites.

6.4.14 Hybrid Seed

Kenya has a hybrid maize programme which is very successful and is relevant in two ways.

1. Peasant farmers are generally considered, with good reason, to be the most conservative group in any society. They know and trust their age-old traditions far beyond the new-fangled, however impressive it may be. They are not prepared to run risks, particularly with their food supply. The hybrid maize programme was successful in that it has been widely accepted by peasant farmers who have now obtained major yield increases and major releases of land for other purposes. This acceptance took a long time, however, and it is this conservatism which indicates how little of the imported, vertically resistant maizes entered the peasant farming gene pool. Such little as did enter it, was by way of accident and was insignificant, and in any event, the peasants had no reason to use it because their own maize was then resistant.

2. Hybrid maizes produced at high altitude are susceptible to *P. polysora* at low altitude. Few data are available but it does appear that these hybrid maizes have a high horizontal resistance to all high altitude diseases but low horizontal resistance to *P. polysora*, which is absent in their centre of breeding. From their resistance to high altitude diseases it must be inferred that comprehensive horizontal resistance is not incompatible with high yield and quality. There is also no apparent reason why low altitude maizes should not be used in a new hybrid seed programme in which their high horizontal resistance to *P. polysora* is retained, in the course of producing massive increases in yield.

6.4.15 Puccinia sorghi

P. sorghi has been present in Africa since the earliest mycological records and, probably, since the original introduction of maize. It has never been a serious disease, at least since scientific records were kept, as the African maizes have high levels of horizontal resistance to it. *P. sorghi* and *P. polysora* are fairly closely related and are both specialised, obligate parasites. It is noteworthy that, in an individual maize host, there could simultaneously be a high horizontal resistance to one of them and a low horizontal resistance to the other. The variable horizontal resistances to these two pathogens appear to be entirely independent of each other, both as survival values and as systems of resistance mechanisms. This emphasises firstly how little we know about the resistance mechanisms themselves and how difficult it can be to obtain mechanistic evidence for the horizontal nature of resistance, and it also emphasises that screening for horizontal resistance to one parasite is unlikely to accumulate horizontal resistance to another parasite.

6.4.16 *Multiplicity of Resistance Mechanisms*

There is little doubt that quantitative horizontal resistance involves many different resistance mechanisms all of which can vary in degree. Collectively, these mechanisms reduce the rates of infection and colonisation of the host and of reproduction in the pathogen. Following the introduction of *P. polysora*, each African maize generation possessed a higher proportion of resistant individuals and the most resistant individuals possessed a higher level of horizontal resistance than in the preceding generation. The same is true at the sub-system level of resistance mechanisms within one host individual. We can only speculate about how many different resistance mechanisms may be involved and such speculation would be irrelevant. What is relevant is that a breeding technique based on the screening for a particular and prominent resistance mechanism is unlikely to accumulate quantitative horizontal resistance. Quantitative horizontal resistance will be maximal only if all resistance mechanisms are present and if all of them are functioning at their highest level, which means that we must breed for horizontal resistance as a single survival value and not for one or another individual resistance mechanism.

6.4.17 *Multiplicity of Parasites*

We have seen that, when *P. polysora* reached Africa, all the major maize parasites were already present. Nevertheless, damaging disease is exceptional in African peasant maize crops, the rule being resistance, and high levels of resistance at that. The accumulation of horizontal resistance to *P. polysora* occurred without any loss of horizontal resistance to any other parasite. In the wide, pathosystem sense, therefore, good horizontal resistance means high levels of resistance to all parasites. It is really quite unimportant which parasite is causing destruction; it is the destruction itself which matters. Indeed, given an unbalanced pathosystem, any minor parasite can cause a major loss, and conversely, given a balanced pathosystem, there is no such thing as a major pest or pathogen. This is what comprehensive horizontal resistance is all about. We must conclude that most of our present crops represent unbalanced pathosystems.

6.4.18 *Multiplicity of Breeding Programmes*

We have seen that each peasant maize crop can be regarded as a separate screening population, each crop being the equivalent of a separate breeding programme except that the control was autonomous rather than determinstic (9.1). The number of peasant maize crops exposed to *P. polysora* is not known, but is probably several millions. Each population was a landrace which was in balance with the local pathosystem. It seems that no one has attempted an exchange of peasant cultivars between various regions in Africa but we can assume it for purposes of discussion. Suppose that maizes were taken from East Africa, which is relatively dry, to West Africa which is relatively wet. Suppose also that a particular leaf blight pathogen is favoured by a wet environment. The East African maizes would then be valueless in West Africa, as each locality has its own pathosystem and the horizontal resistance is in complete balance with one pathosystem only. The conclusion is obvious: breeding for horizontal resistance must not be centralised but must involve a multiplicity of breeding programmes. The equivalent of the millions of peasant crops would be an absurd extreme but the concept of one, central breeding station producing new cultivars on a world-wide basis is no less so.

In this context, mention should be made of an approach which is common among the less experienced overseas aid workers, which involves introducing large numbers of foreign cultivars for trial in a developing country. The workers often have a very understandable preference for the cultivars of their own home country which is temperate and often the least suitable source. This kind of crop introduction is most unlikely to solve any pest or disease problems and even occasionally creates them by introducing susceptibility or by accidently introducing new parasites.

6.4.19 The Number of Host Generations

On a population basis, the decline in the severity of *P. polysora* was due to the fact that each maize generation had a higher proportion of resistant individuals and that the individual extremes of resistance were at a higher level than in the preceding generation. In areas of bimodal rainfall, there are two successive crops each year and adequate levels of horizontal resistance accumulated in 5–7 years, the decline in the severity of damage thus requiring some 10–15 host generations. It should be remembered that this was with a population of extreme initial susceptibility and that, in the later generations, the rate of accumulation of horizontal resistance declined. In some parts of Africa, the rainfall distribution is regular and there is continuous cultivation of maize; the farmer sows a few seeds in a mixed crop every time it rains, the crop thus having a mixture of plants of all ages. Nevertheless, adequate horizontal resistance accumulated in approximately the same period of time in these mixed crops.

These numbers of host generations indicate the time required for breeding programmes. Fewer generations will be needed if the initial host population has more than the unselected level of horizontal resistance and if the intensity of epidemic (or infestation) is artificially maintained as the horizontal resistance increases. On the other hand, more generations will be needed if high domesticated levels of horizontal resistance are required.

6.4.20 The Size of Screening Population

Within limits, the degree of genetic heterogeneity is related to the size of the screening population. In general, each African farm constituted a single screening population of maize, each farmer consistently planting his own seed from his own crop. (More sophisticated farmers might have blamed the cultivar for the high disease levels and discarded their own seed accordingly, which would clearly have been wrong.) The African maize crops were rarely less than one acre or greater than twenty acres. Making allowances for differing host population densities, this means that a one-acre screening population of wheat or rice is probably in the right order of magnitude. This factor is probably not critical although it is clear that this kind of breeding must be conducted on a field scale rather than in a research glasshouse or in small breeders' plots.

6.4.21 The Screening Location

The African maizes illustrate three points about the screening location. (1) As we have seen, the screening location must be closely similar to the environment in which the final selections are to be cultivated. Each area has its own pathosystem and the final selections must be in balance with it. (2) It follows that the screening location may be inconveniently remote from a central research institute; but this is nothing new in agricultural research. (3) When working with temperate crops, it may be possible to use two screening locations, with one in each hemisphere, to double the number of host generations obtained each year. However, this possibility is subject to various difficulties such as seed dormancy, phytosanitation problems, particularly with regard to matching vertical pathotypes, and the similiarity of the two pathosystems.

6.4.22 Selection Pressure for Other Qualities

Insofar as the African farmers kept their best maize cobs for seed, they demonstrated that it is possible to screen for many different variables simultaneously. We have seen that the screening must involve a multiplicity of resistance mechanisms and a multiplicity of parasites. There is no apparent reason why it should not also include a multiplicity of other agricultural qualities. This point is discussed in Chapter 9.

6.4.23 The Negative–Positive Screening Rule

In the African *P. polysora* epidemics, the negative selection pressure for susceptibility was exerted solely by the pathogen. Later, the positive selection pressure for resistance was increased by the farmers who harvested the most resistant cobs, some of which were later used for seed. Natural selection pressures can be increased by deliberate screening. Clearly, there should be a negative screening for the worst individuals before flowering commences, and there should then be a positive screening for all the best individuals before, during or after harvest. The need for such screening may be negligible in the early generations but it will become increasingly important as horizontal resistance accumulates. The positive screening at harvest is likely to provide the best selection pressure for high yields, regardless of the components of that yield.

6.4.24 Seed Screening

When the harvestable product is a seed, such as a cereal or a grain legume, an additional seed screening is possible. This would normally make use of some of the numerous mechanical devices available to seed merchants and seed-testing laboratories. Individual seeds with undesirable characters of any detectable description can then be eliminated. Some of these devices destroy the seed and this is acceptable provided that they only destroy the undesirable seed; for example, soft grains may be undesirable and would be crushed by pressure-controlled rollers which left hard seeds undamaged. Occasionally, destructive tests can be avoided by testing correlated characters. Thus chemical composition may be related to specific gravity. A seed screening can also be used to accumulate horizontal resistance to post-harvest parasites.

Some destructive tests, such as cooking quality, are essential. These can only be conducted on final selections which have been multiplied up as potential new cultivars.

6.4.25 Pure Lines

If an imitation of the African *P. polysora* epidemics is conducted in an inbreeding crop which is to be cultivated as a pure line, the use of male gametocides must cease when adequate horizontal resistance has accumulated. However, the selection pressures must all continue for the 6–7 generations necessary for the establishment of the pure lines. The negative selection pressures will eliminate undesirable segregants and the positive selection pressures will promote desirable lines.

Pollen-cell culture is an alternative method which is still in its infancy but which could lead to considerable saving of time. This technique produces haploid plants which can be doubled to form dihaploids; these are homozygous at all loci.

6.4.26 Host Propagation

There are two aspects of maize propagation which are relevant to breeding other crops for horizontal resistance. Maize has a very high reproductive rate with a multiplication factor of several hundred in about half a year. This is one reason why it is so responsive to selection pressures and breeding of such a crop is both easy and quick compared to perennial crops. Secondly, selection pressures can be negative as well as positive. As we have seen, a maize disease such as streak virus is damaging only sporadically and locally, and more often than not it is absent or unimportant and host erosion of horizontal resistance is then inevitable. This is, perhaps, a rare situation but it does illustrate one of the limitations of horizontal resistance in agriculture. Artificially high levels of horizontal resistance to some parasites may be host eroded in certain crops and in certain farming systems.

6.4.27 Avoidance of Parasite Erosion

If the screening epidemics are conducted with a horizontal pathotype of low horizontal parasitic ability, a subsequent parasite erosion of horizontal resistance can be expected, which is unlikely to be an important factor with obligate parasites but could be a problem with facultative parasites. At its worst, however, it will do no more than delay the completion of the breeding programme. Should polyphyletic pathotypes be troublesome later, the polyphyletic differential interaction (Chap. 7) is now recognisable and a boom-and-bust sequence can be avoided.

6.4.28 *Demonstration of the Horizontal Nature of the Resistance*

No formal attempt has been made to demonstrate the horizontal nature of the resistance to *P. polysora* in the African maizes. The countries concerned can only afford research for the most pressing problems and *P. polysora* is no longer a problem. What we know of it, however, points toward horizontal resistance, the best evidence coming from the quantitative nature of the inheritance, the effects of the resistance and the breeding "method". Vertical resistance could not be found in the African maizes and either does not occur or is very rare. The possibilities of an apparent horizontal resistance due to a multiline effect or a quantitative, incomplete vertical resistance are thus remote beyond serious consideration.

6.4.29 *Measurement of the Resistance*

Equally, no attempt has been made to measure the level of the horizontal resistance to *P. polysora*. We presume that it is at the natural level for that pathosystem, as it controls the disease. Need more be said?

6.4.30 *Prejudice*

We must consider scientific prejudice both past and future. Past prejudice led to the unnecessary and unsuccessful work on vertical resistance to *P. polysora* in Africa. It was the consequence of a dogma, which believed that this was the only resistance available and that all resistance was bound to break down sooner or later. This dogma has affected virtually all plant breeding for disease resistance for several decades.

Future prejudice may be very important, and is a disagreeable topic which must not be evaded. In so far as we can generalise, it is the younger scientists who are most likely to have open minds, to test new ideas and to try new approaches, and the older scientists, often in positions of administrative authority, who are the most likely to oppose them. The opposition will come from scientists who have devoted their careers to a boom-and-bust cycle and who are reluctant to admit that their careers were a waste of time. It will also come from people who have devoted their research to parasite-susceptible cultivars grown with artificial parasite control methods, who have developed those control methods, and who may not relish their results becoming obsolete. Prejudice is an unpleasant failing and it would be pleasing if we could conclude that scientists were free from the baser human emotions. Many of them are, but not all of them. The early developers of horizontal resistance in many crops may have a task not entirely free from hostile interference.

6.4.31 *Pedigree Breeding*

The accumulation of horizontal resistance to *P. polysora* in African maizes was clearly the antithesis of pedigree breeding. With some justification, plant breeders regard themselves as geneticists and want to study genetics. To this end, they keep pedigrees and concentrate on qualitatively inherited characters. Pedigree breeding methods are thus popular, but they are the methods least likely to accumulate horizontal resistance.

Consider the African maizes. There were no controlled crosses and no pedigrees, no measurements or recording of the inherited characters, no "good sources" of resistance, no indentification of resistance mechanisms, no major genes, no tendencies to look backward to the parents of a successful cross, no back-crossing, no vertical resistances and no "race surveys". In a word, there were no plant breeders and no plant pathologists. Such work as was done by breeders and pathologists involved all of these things; it produced vertical resistance and was both fruitless and unnecessary.

We have seen that, in his original theory of evolution, Darwin over-emphasised the individual and under-emphasised the population and the gene pool. His emphasis was at the wrong systems level. This does not mean that the individual must be ignored; all systems levels must be taken into account, but we do obtain a far better comprehension of evolution by studying it at the higher systems level. The same is true of breeding. Pedigree breeders are emphasising the wrong systems level—the individual parent, the individual character and the individual gene. Population breeders operate at a higher systems level. They exert many selection pressures on a flexible gene pool and, in so doing, stand a better chance of obtaining a balanced system and domestication. This point is illustrated when we speak of "population breeding" instead of "plant breeding".

6.4.32 *Recapitulation*

We can now summarise the factors which are necessary for the successful accumulation of horizontal resistance.

There is no need to search for a source of resistance and even the most susceptible cultivars will accumulate horizontal resistance. It is essential to eliminate vertical resistance or its effects. There should be random cross-pollination in a genetically flexible population but the genetic base can probably be quite narrow. Selection pressure is exerted by the parasite and should occur as naturally as possible. The screening should lead to an increase in all variable resistance mechanisms against all locally important parasites; for this reason there should be a breeding programme for each ecological zone. Some 10–15 host generations are necessary in a population of some thousands of individuals which must be screened in the area of their future cultivation. Selection pressure for all agronomic qualities must be exerted by using negative–positive screening techniques; seed screening may also be possible and destructive screening may be conducted on bulked lines. Care should be taken to avoid both host erosion and parasite erosion of horizontal resistance after the completion of screening, and demonstration and measurement are inherent in the technique. Scientific prejudice is likely to be the greatest hindrance to work on horizontal resistance. These techniques are the antithesis of pedigree breeding.

6.5 Annual Crops

The accumulation of horizontal resistance in the African maizes can be imitated in other crops and we must now examine some examples. If this imitation is to be

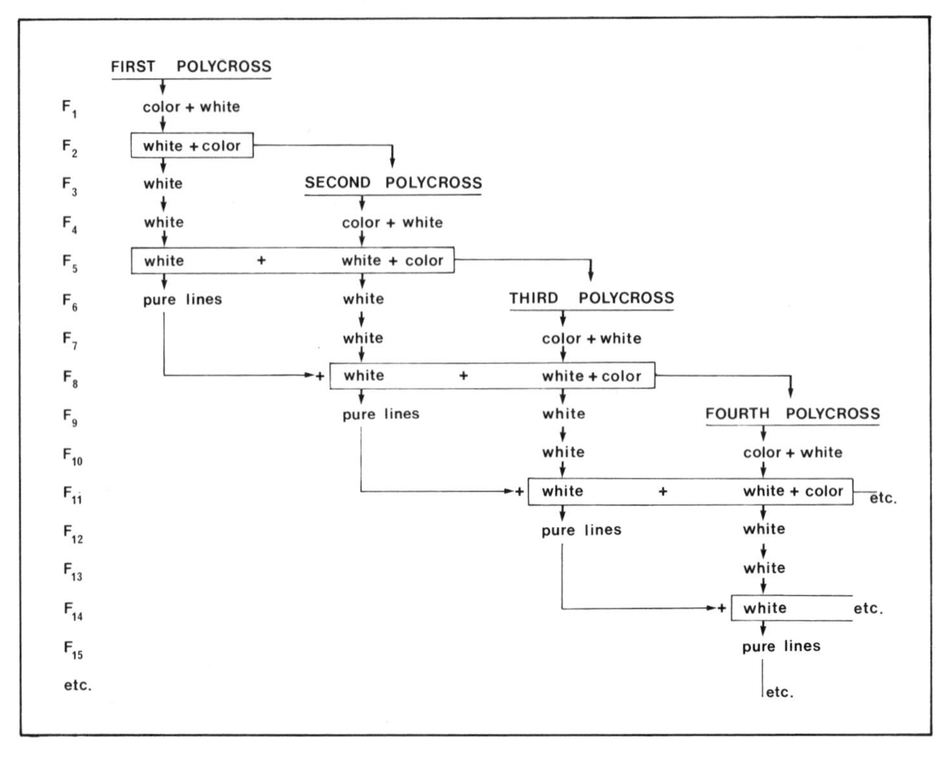

Fig. 13. The Stoetzer strategy

reduced to the most simple terms, it involves the avoidance of vertical resistance and the avoidance of all other features of pedigree breeding.

6.5.1 *The Awassa Bean Programme*

A deliberate attempt is being made to imitate the African maizes in haricot beans *(P. vulgaris)* at Awassa, Ethiopia (Stoetzer, 1975). The Stoetzer strategy is summarised in Figure 13. The programme is designed to produce new cultivars of white pea beans with comprehensive horizontal resistance to all locally important parasites. More than 100 cultivars of white beans, of various shapes and sizes, were mixed and sown in rows as female parents. A mixture of some sixty cultivars of coloured beans was sown in alternate rows to act as male parents. The F_1 was sown exclusively from white bean parents and produced a proportion of coloured beans indicating a cross-pollination of 5%. The F_2 was sown exclusively from the coloured beans of F_1 and produced white segregants which are then grown for a further four generations to produce pure lines. Both white and coloured beans of the F_2 are used for a second random polycross and so on.

As far as could be ascertained, all the parents were susceptible to all the locally important parasites; if vertical resistance was present, it was not operating. In each generation, there is a negative screening just before flowering—the least healthy 30% of individuals are removed. There is also a positive screening at harvest—the most healthy individuals being retained. There is also a seed screening on the basis of seed characters. Various details designed to intensify the natural epidemics and infestations can be assumed.

This program was started in 1974 and the climate permits two host generations each year. The results will be published in due course. Stoetzer (1975) is also conducting similar programmes with tomatoes *(Lycopersicon esculentum)* and chillies (*Capsicum* spp). However, the latter crop can only be cultivated at a rate of one generation each year.

6.5.2 The Mexican Potato Programme

J. S. Niederhauser, working on potatoes in the Toluca Valley of Mexico, should be regarded as the pioneer who, having recognised the inadequacies of vertical resistance, deliberately set out to breed for horizontal resistance. This area is the centre of origin of *P. infestans* and functional oospores are common, due to the presence of both mating types. As a result, a wide range of vertical pathotypes occurs and vertical resistance breaks down so quickly that it is agriculturally valueless. The esodemic is also particularly severe because the pathogen is in a state of perfect balance with the local climate. High levels of horizontal resistance to *P. infestans* have been obtained.

6.5.3 The Kenya Potato Programme

This programme was begun in 1968 and the results obtained by mid-1972 have been described (Robinson, 1973b).

Potatoes in Kenya should be compared with maize in Kenya rather than with potatoes in Europe. That is, they are essentially a source of abundant, cheap food in a country with one of the highest population growth rates in the world whose people are largely engaged in peasant farming on a semi-subsistence basis. However, if potatoes are to become this, they must be grown by small holders who know how to grow maize, which means that the prevention of crop loss due to parasites must cost the farmer nothing, either in effort, or technical expertise, or in money spent on fungicide spraying and routine use of expensive, certified seed tubers. Kenya has some six million acres of high-potential land which is at too high an altitude for the successful cultivation of maize as a subsistence crop, but is however ideal for potatoes which would constitute a fully acceptable, alternative, staple diet if their cultivation was both cheap and easy. The successful development of comprehensive horizontal resistance in potatoes would increase the total amount of high-potential land available to peasant farmers in Kenya by 30%.

In practice, high levels of horizontal resistance must be accumulated to only two major pathogens. These are blight *(P. infestans)* and bacterial wilt *(Pseudomonas solanacearum)* and the task is facilitated by Kenya's short-day climate. Other tuber-borne diseases, particularly the temperate viruses, are not serious. Healthy plants can be found in old cultivars which have been grown for seventy vegetative generations without renewal of stocks, so that peasant farmers can produce their own seed tubers simply by harvesting their most healthy plants separately for seed, and need purchase certified seed only if they want a new cultivar.

Because one potato berry produces many seeds, all cross-pollination was done manually and it is possible to produce seeds and screen seedlings at an average rate of one thousand each working day. It is thus possible to screen 150,000 seedlings per generation, with two generations each year. Scions of the most promising survivors are grafted to tomatoes for production of true seed and rapid initiation of the next generation. Vertical resistance to blight is eliminated epidemiologically and vertical resistance to bacterial wilt apparently does not occur, (Robinson, 1973a).

Dickinson (1972) developed a stem-cutting technique for the rapid multiplication of new cultivars and Bruce (1973) showed that six tubers can be multiplied to five tons of tubers in 180 days, so that new cultivars can thus be distributed in relatively small quantities to many individual farmers very rapidly.

6.5.4 The Scottish Potato Programme

This programme was designed by Simmonds (1966) to resynthesise tuberosum potatoes from andigena potatoes by population breeding methods, and commenced in 1959. It is unquestionably the first attempt to move away from pedigree breeding methods in potatoes, and can usefully be compared with the Hawaiian work on sugarcane (p. 117). The main objectives are to produce the long-day characteristic and horizontal resistance to blight. Vertical resistance to blight does not occur in andigena potatoes. Considerable population responses to these selection pressures have been demonstrated. The only criticism of this immensely important initiative is that it produces one generation in only two years.

6.5.5 Wheat

With the possible exception of tomato breeders, the wheat breeders constitute the strongest bastion of the vertical resistance dogma and little work has been done on horizontal resistance in this crop. The most important results are those of Lupton and Johnson (1970) who showed how great the vertifolia effect is in modern cultivars with vertical resistance to *P. striiformis*, and also that the high horizontal resistance of "Little Joss" was not correlated with its detrimental long straw and that high levels of horizontal resistance can be combined with other desirable agronomic characters.

Robinson (1973a) proposed a breeding strategy for wheat based on an imitation of the African maizes. The first step is the epidemiological elimination of the effects of vertical resistance followed by a random polycross assisted by the use of male gameticides. Marker genes could be used if desired and a Stoetzer strategy is feasible.

Mention should be made of composite crosses, which involve the crossing of some thirty different wheats in all combinations and their subsequent cultivation for several generations as a segregating, mixed population, a technique which comes very close to accumulating horizontal resistance, and in the past only failed to do so because vertical resistance, or its effects, was not eliminated. Furthermore, after the first crosses had been made, the amount of crossing in subsequent generations was at a low level.

6.6 Sugarcane

In his 1936 presidential address to the American Phytopathological Society, G. H. Coons (1937) spoke about controlling plant diseases by the development of resistant varieties. He described how the historic sugar industry of Louisiana was virtually bankrupted in the years 1923–27 due to the three major diseases mosaic, red rot and root rot, diseases to which all the old noble canes were highly susceptible. They were replaced with new, resistant varieties some fifty years ago, and the Louisiana State averages of sugar yield increased by at least 50%, and have never subsequently declined below this level, in spite of incorrect prognoses of a boom-and-bust cycle (see below). The remarkable feature of this situation is not that the problem was solved, but that, being solved, it was forgotten. These sugarcane diseases have now been unimportant for many years and, being unimportant, they do not feature prominently in plant pathological teaching. But they should do so, because, from a teaching point of view, they are an admirable illustration of how a virtually disease-free agriculture can be achieved. In a word, they illustrate that disease resistance can be permanent.

The cane breeders were lucky. No doubt, they would have exploited the easy breeding methods which result from oligogenically inherited resistance, had they been able to find such resistance, but they could not find it because it does not occur.

6.6.1 The Absence of Vertical Resistance

The progenitors of sugarcane are perennial, evergreen, tropical grasses with a natural vegetative propagation, which exhibit great spatial and sequential continuity of host tissue. As a result, the esodemic is continuous and vertical resistance cannot evolve because it has no survival value. It can break down but it cannot recover. It seems that oligogenically inherited resistance has never been recorded. Nevertheless, vertical resistance is frequently assumed and has often been postulated.

6.6.2 Apparent Vertical Resistance

Vertical resistance to red rot (*Glomerella tucumanensis*) was postulated by Abbott (1938, 1961). The cultivar POJ 213 was introduced to Louisiana in the 1920's and was considered highly resistant when it was released to growers. It had four years of relative freedom from red rot but, in 1930, it suffered severely and, by 1934, had ceased to be a commercial variety. It appears that the freedom from red rot was due to the release of disease-free seed setts, combined with a high leaf resistance and low stem resistance. Leaf lesions are the main source of inoculum for stem infection and the effect of the disease-free setts was thus enhanced. However, Abbot (1938) demonstrated "physiologic specialisation" with an impressive array of inoculation experiments. It is now clear that this differential interaction between pathodemes and pathotypes was not due to vertical resistance and was probably a polyphyletic pathosystem (Chap. 7). Chona and Padwick (1942) recorded a similar phenomenon in India.

There is also a differential interaction between cane pathodemes and mosaic virus pathotypes which are differentiated with the cultivars CP.29/291, CP.31/294 and Co.281 (Abbott, 1961). This again appears to be a polyphyletic differential interaction. Geographical differential interactions have been recorded with ratoon stunting but are apparently due to environmental factors as this disease is only damaging when the host is suffering from water stress, to which some cultivars are more sensitive than others. At least one geographical differential interaction with sugarcane smut has been shown to be due to a mixture of cultivars, and mislabelling of cultivars can also occur. The "impermanent nature" of the resistance to Pokkah Boeng (*Gibberella moniliforme*) has been postulated (Martin *et al.*, 1961) but apparently only on the evidence of the great variability of this taxonomically confused fungus. There are also two rusts of sugarcane (*Puccinia erianthi* and *Puccinia kuehnii*) for which vertical resistance has been postulated solely on the grounds that they are rusts.

It should be added that horizontal resistance occurs against all these pathogens, as well as against some twenty other major pathogens of sugarcane. Relatively few cultivars have comprehensive horizontal resistance, however, and this must be regarded as a failure of some seventy years of cane breeding. But cultivar resistance has proved permanent, and the relatively few resurgences of a disease such as mosaic are due to carelessness and the fact that new cultivars were released after inadequate testing and subsequently proved to be susceptible.

6.6.3 Traditional Cane Breeding

Like the traditional potato breeders, cane breeders had used pedigree breeding methods. This is a pity, as they could have been blazing a new trail for other breeders to follow, but their thinking has been dominated by pedigrees—the parentage of good cultivars. Pedigree breeders tend to look backwards to the parents, and this is the very reverse of evolution. Population breeders look forward; they are exclusively interested in progenies, and this is evolution; a process of change in which the fittest survive.

Pedigree breeding has other disadvantages. It is labour-intensive and requires much labelling, measuring and recording. This sets severe limits both to the number of selection criteria which can be employed and to the numbers of individuals which can be screened. It is difficult to obtain comprehensive resistance this way and pedigree breeding shows a strong tendency to select for resistance to only one or two parasites at a time. The cane breeders have also tended to employ totally inappropriate, artificial inoculation techniques. These often involve wounding, exorbitantly high dosages injected under pressure, and so on. They have also tended to rely far too heavily on foreign cultivars selected for a very different pathosystem. Finally, cane breeders have allowed themselves to be quite unnecessarily influenced by fears of a boom-and-bust cycle, when they could and should have demonstrated the nature and value of horizontal resistance to all other plant breeders. Instead, they blindly followed an absurd fashion and concluded that all resistance, in all plants, including sugarcane, was bound to breakdown sooner or later. They were well placed to have been the leaders of a different and better fashion.

6.6.4 *Cane Breeding in Hawaii*

Cane breeding in Hawaii is very different. It involves a random polycross technique called the "melting pot", using good cultivars as parents. Whereas the traditional, pedigree cane breeders are limited to about 100,000 seedlings each year, the Hawaiian breeders screen three million, using a graded series of screening tests with the easiest first and the most complex last. One weakness of this system was the tendency for their cane varieties to be very susceptible to smut *(U. scitaminea)* which was absent from Hawaii. When smut was accidentally introduced, the breeding techniques were justified and resistant varieties were quickly developed.

As we have seen, cane growers in Hawaii enjoy a cane culture which is virtually pest-free and disease-free, a freedom from crop loss due to parasites which costs the farmers nothing and is permanent. Not only are Hawaiian cane and sucrose yields the highest in the world, but it is also noteworthy that, although the Hawaiian canes are a balanced pathosystem in Hawaii, they are not balanced in other parts of the world, where they generally prove suceptible to one parasite or another.

This Hawaiian cane breeding has been in progress for many years but all other cane breeding stations have tended to remain traditional, as have the potato breeders, with the notable exception of Simmonds. There appears to be no good reason why the example of the Hawaiian cane breeders should not be followed in all crops.

6.7 Long-Term Perennial Crops

6.7.1 *Difficulties*

Perennial crops are difficult to breed for several reasons. In the natural, autonomous, evolutionary system, changes occur during periods of geological time and the difference between one hundred days and one hundred years in an individual life span is then of no significance. But in the artificial, deterministic, evolutionary system, changes occur during periods of historical time, when differences in life span are crucial. In the tropics, for example, it is possible to obtain three bean generations in one year, but it is not possible to obtain more than one coffee generation in three years, and the generation time in coconuts is six years.

There are also major differences in the numbers of seeds produced. One coffee tree will produce several thousand seeds after three years; one coconut palm will produce about sixty seeds after six years. The size of the individual plant is another problem. An acre or less of wheat or rice will provide a screening population of adequate size and heterogeneity, while a comparable population of coffee would occupy about one thousand times this area, and coconuts would require about ten thousand acres. There are other problems also, associated with difficulties in cross-pollination, the replanting of successive generations and so on. Theoretically, it is possible to imitate the African maizes with a perennial crop but, in practice, it would be difficult, slow and expensive. We must consequently look for short cuts.

6.7.2 *Exploitation of Existing Host Populations*

Perhaps the final lesson we can learn from the African maizes is the value of existing, cultivated crops as screening populations. This approach will overcome many of the difficulties listed above for perennial crops and is additionally valuable for several reasons.

Most perennials are outbreeders for obvious, evolutionary reasons. Peach and arabica coffee are notable exceptions but even arabica coffee has about 3% of outcrossing and shows great heterogeneity in Ethiopia which is its centre of diversification.

Most perennial crops are less domesticated than annual crops, which means that primitive cultivars and even wild progenitors are more likely to have agricultural value than those of annual crops. Individuals selected for resistance from a heterogeneous population may thus be of immediate utility. Grapes, citrus and the stone and pome fruits are the obvious exceptions to this rule, but these exceptions must be equated with the innumerable forest tree species, tropical plantation tree crops and so on.

Vertical resistance is rare in perennials and if it does occur, it is likely to be associated with specialised features such as a pathogen-induced leaf fall and it can usually be recognised and avoided. This approach thus favours horizontal resistance.

Lastly, most perennials can be propagated vegetatively and high levels of horizontal resistance, as well as other agricultural qualities, can be preserved without difficulty. Coconut is the exception which proves the generality of this rule. In some perennials, vegetative propagation also increases the rate of multiplication. This is not true of crops with dimorphic branching, however. In cocoa and coffee, cuttings can only be taken from orthotropic branches and the rate of vegetative reproduction is only a small fraction of the seed reproduction rate.

We must now consider some examples.

6.7.3 *Blister Blight of Tea*

Cultivated tea is a freely intercrossing hybrid spectrum between the two extremes of *Thea sinensis* and *Thea assamensis*. Traditionally, tea is propagated from seed and, in such crops, the variation is so great that it is doubtful if any two tea bushes in the world are identical. In tea crops propagated from seed a large proportion of the yield consequently comes from a small proportion of the bushes and the overall quality is low due to both the mixture of inherent qualities and the variation in fermentation times. Commercial plantations of seed-propagated tea have been screened for promising individuals. The screening is divided into several stages with a drastic reduction in the numbers of individuals and a considerable increase in the complexity of the tests in each stage. An original population of many thousands of individuals ultimately yields only a few individuals suitable for vegetative propagation as clones. The yield and quality of these clones are at least double and possibly five times as great as those of seed-propagated teas.

Tea thus provides an excellent example of the use of existing, cultivated crops as screening populations, as any screening criteria can be used and disease resistance is no more difficult to select than, say, yield. Blister blight, caused by *Exobas-*

idium vexans, is a leaf disease which destroys the harvestable product. It is normally controlled by spraying with a fungicide in accordance with a well-tested forecasting system based on sunshine hours. Because a tea crop can be harvested for about one hundred years before replanting, there is a natural reluctance to replant existing tea solely for reasons of blister blight control. However, there is no reason why blister blight resistance should not be included in the selection criteria for new clones, at least in areas where blister blight occurs. It is not known whether vertical resistance occurs in tea but even if its occurrence is improbable, work on resistance to blister blight should embrace evidence for the horizontal nature of that resistance. Polyphyletic pathotypes and pathodemes may occur (Chap. 7). Finally, any screening of seed-propagated tea populations should obviously be conducted in the absence of fungicidal treatments.

6.7.4 *Coffee Berry Disease*

Coffea arabica is an allotetraploid derived from two wild diploid species. The diploid progenitors are believed to be the East African species *Coffea eugenioides* and one of the West African species, most probably *Coffea canephora* (robusta coffee). The distribution of these species overlaps in Uganda which is the most likely centre of origin. As often happens with allotetraploids, the centre of diversification is different from the centre of origin and is in Southwest Ethiopia where the diploids do not occur. *C. arabica* later became extinct in the centre of origin where the climate is unsuitable.

Although it is an old-world crop, coffee is notable in that it is not mentioned in biblical or ancient Egyptian, Greek and Roman records. In a study of the Roman spice trade, Miller (1969) showed that cinnamon (*Cinnamonum* spp.), which then occurred only in southeast Asia, was shipped across the Indian Ocean by Indonesians to Madagascar. It was subsequently shipped to the East African coast and taken by caravan to southwest Ethiopia. The caravan route then branched, with one route to the Nile and one to the Red Sea. These traders thus traversed the centre of diversification of *C. arabica* and, given the sophistication of the Roman spice trade, it was unlikely that *C. arabica* was present in the area at that time. The first record of coffee is an Arabian one of 850 AD and it is probable that *C. arabica* originated around 600 AD.

All the arabica coffee of the world is derived from the Ethiopian material. It was taken to Arabia in the 14th century; to Java and Reunion by the Dutch and French respectively in the 16th century, later to Europe and then to the New World. Each shipment was made with seed which remains viable for only a few months. With each shipment there was a progressive narrowing of the genetic base and much of the New World coffee is derived from a single tree grown at Versailles. Finally, the British took narrow gene base coffee to Kenya and, after World War I, launched a major coffee expansion programme in Western Kenya. *C. arabica* came into contact with its diploid progenitors for the first time in some fourteen centuries and proved to be highly susceptible to a new disease called coffee berry disease (CBD), caused by *Colletotrichum coffeanum* (MacDonald, 1926).

C. coffeanum is a micro-epiphyte of coffee bark, probably of *C. eugenioides*. It occurs in various forms which are distinguishable in culture. It has water-borne spores and the natural allo-infection is probably a passive transmission by birds and insects. However, two of the forms possess the additional dispersal mechanism of invading the pericarp of coffee berries. One of these forms causes the relatively harmless disease called brown blight and was apparently taken with *C. arabica* to Ethiopia and to all subsequent coffee areas. The brown blight pathosystem has remained balanced. The other form is CBD which was apparently left behind in the centre of origin when coffee was taken to Ethiopia. In the course of some fourteen centuries of cultivation in the absence of this parasite, there has been a major host erosion of horizontal resistance and all the arabica coffee of the world is highly susceptible to it. In 1971, CBD was recorded in Ethiopia for the first time, and it soon became obvious that the Ethiopian coffee crop was facing ruin.

CBD is a typical anthracnose which attacks the green berry. Although the Ethiopian coffee crops are highly susceptible on a population basis, there is wide variation between individuals, the most susceptible individuals suffering a complete loss of berry early in the season, while the most resistant remain free from the disease at harvest. Robinson (1974) launched a selection programme for resistant individuals. Resistant trees occur at the frequency of 0.1–1.0%; they are identified and the first harvest is kept for seed. During the three years which the progeny require to come into bearing, the parent tree is repeatedly tested for resistance, yield and quality. The progeny is tested for homozygosity and resistance. On the basis of these tests it is intended to keep the best 10% of selections and a target of 500 selections was set; four hundred selections had been made by the end of 1974. The programme is now being conducted and refined by van der Graaff (1975) who expects to have 50 new, CBD-resistant cultivars, each tested for resistance, yield, quality and homozygosity, and each consisting of about 1000 trees bearing seed, within seven years of the programme being initiated.

It should be added that Ethiopian coffee is cultivated according to centuries old traditions, with a random spacing of unpruned, mixed populations whose yields and quality are modal. The appearance of CBD now compels replanting with resistant trees and consititutes a unique opportunity for the modernisation of coffee culture. A major economic loss can thus be converted into a major economic gain.

6.7.5 Bayoud Disease of Dates

A good date palm *(Phoenix dactylifera)* will yield some 400 kg of dates each year for about 70 years. Dates are ideally suited to a desert climate provided that they can be irrigated, and are the staple food and the main source of wealth in oases throughout the Sahara and Arabian deserts. The date palm does not breed true, the best dates being cultivated as clones and muliplied by the suckers produced at the base of the palm at an average rate of four each year. Seed-propagated dates are immensely variable and normally produce fruit fit only for feeding camels.

Bayoud disease, caused by *F. oxysporum* f.sp. *albidinis* first appeared in Morocco late in the last century, at which time, Morocco had fifteen million date palms; ten million high quality, clonal dates and five million seed-propagated and relatively valueless. The ten million clonal dates are now dead as a result of bayoud disease which has since spread into Algeria. There seems to be no way of preventing its ultimate spread to all the desert regions of North Africa and Arabia. The only feasible control is by the development of resistant varieties and clearly the resistance must be horizontal. However, dates are particularly difficult to breed, being dioecious and having a generation time of 7–10 years. The only quick answer to the problem of Bayoud disease seems to lie in the screening of Morocco's five million seed-propagated palms for reasonable yield and quality and to test the selections for resistance. Such a programme is now being conducted by a French team at Zagora, in Morocco. The final selections may be inferior to the lost clones but they will be preferable to no dates at all. However, vegetative propagation of dates is also a very lengthy business, and it is estimated that it would take 150 years to replace the ten million lost palms from one resistant individual, using the traditional propagating techniques. Research is thus necessary on more rapid multiplication techniques.

6.7.6 *Cadang-cadang of Coconuts*

Cadang-cadang is a lethal disease of coconuts which is thought to be a virus and which occurs only in the Philippines, where it is estimated that, by 1972, the disease had killed some twelve million palms and caused a loss of U.S. $ 250 million. The disease spreads at a rate about three miles each year and is leaving a wake of dead palms in the previously flourishing coconut industry of Luzon. There is no known control except that palms are usually killed in middle age and the use of high-yielding hybrids with an increased replanting rate may be economically justified in spite of the disease.

The disease first appeared some fifty years ago on San Miguel Island, on the eastern coast of Luzon, and out of an original palm population of one quarter of a million on the Island, only 80 palms have survived. Resistance is apparently manifested by the age of death; resistant individuals are killed more slowly and have a longer, economic life. This resistance is poorly understood but, within these limits of knowledge, it has the characteristics of horizontal resistance. If resistance to cadang-cadang is to be found, it will clearly be most easily and quickly discovered among the 80 San Miguel survivors, which represent 0.032% of a population which has been exposed to the disease for 50 years and in which all other individuals have proved susceptible and are dead. The chance of any one of these survivors being an accidental escape from infection is remote. This is an intensity of natural selection which would be impossible to achieve either quickly or cheaply in a formal breeding programme. The 80 survivors, and other survivors on the nearby mainland, may thus be one of the coconut industry's most valuable possessions. Bigornia (1972) has established resistant lines which are now being used to replant ruined farms.

6.7.7 *Stone and Pome Fruits*

The stone and pome fruits have undergone domestication for several millenia. They are propagated vegetatively and genotypes which are intermediate between the wild progenitors and modern cultivars are lost. With these crops, there seems to be no alternative to the difficult and laborious creation of screening populations.

6.7.8 Bananas

The edible banana is a sterile triploid. It is thus one of the most difficult crops to breed. Its centre of origin is in Southeast Asia but it is now extensively cultivated in central and South America which is the centre of origin of *P. solanacearum*, the cause of Moko disease. In view of the separate evolution of host and parasite and the difficulty of breeding the host, Moko disease can be quoted as possibly the most difficult of resistance-breeding problems.

Chapter 7 Polyphyletic Pathosystems

The word polyphyletic means that a plant has a multiple origin, which may be qualitative, as with an allopolyploid, or quantitative, as with a fully fertile hybrid swarm showing all differences in degree of hybridisation between different species.

The polyphyletic pathosystem is a super-system; it is a pattern of sub-systems and each sub-system is an ordinary horizontal pathosystem.

7.1 The Qualitative Polyphyletic Pathosystem

7.1.1 Qualitative Polyphyletic Origins

At its most simple, a qualitative polyphyletic origin involves the hybridisation of two diploid species of equal chromosome number. The hybrid chromosome number then doubles to form an allotetraploid. *C. arabica* is such an allotetraploid ($2n=44$) derived from two wild diploid species ($2n=22$).

Allopolyploids are frequently more complex than this. The chromosome contributions from the original progenitors may be unequal in number, more than two progenitors may be involved, as with wheat, or there may be further hybridisation as with the wheat × rye hybrid Triticale, and individual chromosomes may be gained or lost to form aneuploids. With such complicated genomes the polyphyletic pathosystem is correspondingly complex but the basic principles remain unaltered. We shall now consider a simple, hypothetical, qualitative, polyphyletic pathosystem.

7.1.2 The Simple, Qualitative, Polyphyletic Pathosystem

This pathosystem is illustrated in Figure 14. The two host progenitors (A and B) are different species of one genus and have the same chromosome number. The hybrid is infertile at the diploid level and is a fully fertile, simple allotetraploid. Each diploid progenitor is susceptible to its own *forma specialis* of the one parasite species but has complete polyphyletic resistance (7.2) to the *forma specialis* of the other progenitor species. For simplicity, we assume that there are no vertical genes in the pathosystem and we measure damage on a scale 0–4; we also assume that the parasite *formae speciales* can hybridise. Let us now examine the properties of this pathosystem.

7.1.3 The Context of Host-Parasite Interaction

We can refer to the various hosts of Figure 14 as polyphyletic pathodemes, and to the various parasites as polyphyletic pathotypes.

	Host Species A	Hybrid AB	Host Species B
Parasite f.sp. a	4	2	0
Hybrid ab	2	4	2
Parasite f.sp. b	0	2	4

Fig. 14. The qualitative polyphletic pathosystem

The first and obvious property of the pathosystem is that there is "physiologic specialisation", as it used to be called. That is, there is a differential interaction between polyphyletic pathodemes and polyphyletic pathotypes. But this differential interaction is not due to a vertical pathosystem; it is demonstrably not a Person differential interaction. We must note that this is a qualitative system in terms of the hybridisation of pathodemes and pathotypes, but that the interaction is essentially quantitative in that it is due to differing amounts of polygenically inherited, horizontal resistance and parasitic ability.

7.1.4 The Context of Systems Level

Let us assume that the host allotetraploid is of fairly ancient origin and that there has been enough time for considerable mutation and segregation of polygenically inherited characters, and the same for the parasite hybrid. Let us consider only the one interaction, between the host hybrid and the parasite hybrid.

The host is a population in which there is a normal distribution of all variable survival values, including polygenically inherited horizontal resistance. The parasite is also a population with a normal distribution of horizontal parasitic ability. This one interaction is thus a complete horizontal pathosystem according to the descriptions used so far in this book, and is represented by the Person model (Fig. 11). The same is true for any one of the interactions of the polyphyletic pathosystem. The polyphyletic pathosystem is thus a super-system of different but related horizontal pathosystems.

7.1.5 Natural Parasite Hybrids

Let us assume that host species A and B are both present in a natural pathosystem, and that f.spp. a and b are also both present. We can presume that the *formae speciales* hybridise naturally with a low but nevertheless fixed frequency. This is part of the natural variability of the parasite population and is a feature of its micro-evolutionary changes.

Let us assume also that damage level 4 in Figure 14 is the natural level of damage and is such that the micro-evolutionary survival of the host is not impaired. It is the same level as level 5 in the Person model (Fig. 11). So long as the host hybrid is absent, the parasite hybrid will become extinct as frequently as it is formed, because it can only produce damage level 2 on host A and host B.

7.1.6 *Parasite Erosion of Horizontal Resistance*

If plant breeders produce a new crop plant by the formation of an interspecific or intergeneric allopolyploid, it will have an apparently high level of horizontal resistance to all the parasites of either host progenitor. This apparently high level of resistance will be reduced by the hybridisation of any of the parasites and this is a form of parasite erosion of horizontal resistance.

7.1.7 *Artificial Parasite Hybrids*

It is often postulated that experiments in the artificial hybridisation of plant parasites are dangerous and, equally romantically, that such work forms the basis of any biological warfare involving crops. This is only true if existing crops are already vulnerable (p. 131); if there is no crop vulnerability, parasite hybridisation can only reduce horizontal parasitic ability.

7.1.8 *The Context of the Vertical Pathosystem*

The polyphyletic pathosystem has some of the features of the vertical pathosystem and confusion between the two is possible. As we have seen, there is a differential interaction but the polyphyletic differential interaction is very different from the Person differential interaction.

Secondly, we defined vertical resistance as being conferred by mechanisms which are within the capacity for micro-evolutionary change of the parasite. At first sight this is true of the polyphyletic pathosystem, but there is an important difference. The breakdown of vertical resistance is the failure of a resistance mechanism which is actually present in the host, while the parasite erosion of the horizontal resistance of a polyphyletic pathodeme involves resistance mechanisms which are absent. It is the loss of an apparent resistance which was never present; this is the manifestation of vulnerability (8.1).

Thirdly, in an agricultural context, vertical resistance is temporary resistance which leads to a boom-and-bust cycle of cultivar production, one of the most important features of vertical resistance in terms of practical pathosystem management. With poor management of the vertical pathosystem, this cycle is repetitive and unending. But the loss of apparent horizontal resistance in a qualitative, polyphyletic hybrid is quite different, because it happens only once, and is neither repetitive nor a cycle.

Let us consider an example. The wheat-rye hybrid called Triticale is resistant to *P. graminis (sensu ampliss)*. That is, both *P. graminis triticale* and *P. graminis secalis* have low horizontal pathogenicities to it. Given an appropriate degree of hybridisation between these two *formae speciales*, the apparent horizontal resistance of *Triticale* to stem rust will be lost, and the level of the true horizontal resistance will be revealed. This level will depend on the levels of horizontal resistance in the two progenitors to their respective *formae speciales;* but the

Triticale pathosystem is a horizontal sub-system within the polyphyletic super-system. The level of horizontal resistance to the new polyphyletic pathotype can be increased to an adequate level within the Triticale sub-system; it is horizontal resistance.

Finally, in Figure 14 we assumed, for the sake of simplicity, that no vertical genes were present. In practice, of course, vertical genes could easily be present, which would make resolution of the entire pathosystem much more difficult but not beyond elucidation, either theoretically or experimentally.

7.2 The Quantitative Polyphyletic Pathosystem

7.2.1 Quantitative Polyphyletic Origins

At its most simple, a quantitative polyphyletic origin involves the hybridisation between two inter-fertile species with subsequent back-crossing and inter-crossing to form a spectrum of all degrees of hybridisation between the original progenitors. Cultivated tea constitutes such a spectrum between two progenitor species. Other quantitative polyphyletic origins are more complicated. Modern cultivars of sugarcane, for example, involve varying degrees of hybridisation of noble cane *(Saccharum officianarum)* with the two wild species *Saccharum robustum* and *Saccharum spontaneum*.

A model of a simple, quantitative, polyphyletic pathosystem is shown in Figure 15 and it will be seen that it has no fundamental differences from the qualitative polyphyletic pathosystem (Fig. 14).

		A ← Host Hybrid Spectrum → B									
		8:0	7:1	6:2	5:3	4:4	3:5	2:6	1:7	0:8	Sum
a ← Parasite Hybrid Spectrum → b	8:0	8	7	6	5	4	3	2	1	0	36
	7:1	7	8	7	6	5	4	3	2	1	43
	6:2	6	7	8	7	6	5	4	3	2	48
	5:3	5	6	7	8	7	6	5	4	3	51
	4:4	4	5	6	7	8	7	6	5	4	52
	3:5	3	4	5	6	7	8	7	6	5	51
	2:6	2	3	4	5	6	7	8	7	6	48
	1:7	1	2	3	4	5	6	7	8	7	43
	0:8	0	1	2	3	4	5	6	7	8	36
	Sum	36	43	48	51	52	51	48	43	36	

Fig. 15. The quantitative polyphyletic pathosystem

7.2.2 The Super-system

The matrix of Figure 15 is the interaction of a host hybrid spectrum with a parasite hybrid spectrum. It is a super-system of 81 horizontal sub-systems. Each sub-system is a Person model but, for purposes of discussion, we can regard each sub-system as being effectively non-variable. Let us also regard the super-system as a single, natural pathosystem. We assume that the complete hybrid spectrum occurs in nature, with both the host and the parasite. We assume also that 8 is the natural level of parasite damage. It will be observed that no amount of parasite hybridisation will raise the parasite damage above the natural level, but will always decrease the parasite damage except when the degree of parasite hybridisation equals the degree of host hybridisation. Host hybridisation also decreases the amount of parasite damage, except when the degree of host hybridisation equals the degree of parasite hybridisation. At first sight, hybridisation is an advantage, in that a natural pathosystem corresponding to the matrix of Figure 15 would suffer less parasite damage than a pathosystem consisting uniformly of the same degree of hybridisation of both host and parasite.

If we now examine the total parasite damage of all columns and all ranks of the matrix in Figure 15, we see that there is least overall damage with least hybridisation in either the host or the parasite; and most overall damage occurs in the 4:4 host hybrid, and most damage is caused by the 4:4 parasite hybrid. A natural pathosystem which was identical to the super-system of Figure 15 would thus be an unbalanced system. The 4:4 hybrids would suffer and cause more parasite damage, on average, than the unhybridised progenitors. The hybrids would tend to be lost due to negative selection pressure, and we must conclude that such a super-system would be rare in nature.

Finally, if this polyphyletic super-system occurred naturally, there would be a mixture of pathodemes and a mixture of pathotypes, a situation it would be tempting to equate with a vertical pathosystem. This is yet another example of apparent vertical resistance and an indication that horizontal resistance can reduce the exodemic as well as the esodemic.

7.2.3 Polyphyletic Resistance and Parasitic Ability

The relationship of immunity to horizontal resistance has already been discussed. Absolute horizontal resistance can be eroded; immunity cannot be eroded. However, recognition of the polyphyletic pathosystem compels us to modify these definitions. We must recognise two categories of immunity which either can or cannot be diluted by hybridisation. Wheat is immune to coffee rust and this immunity cannot be diluted because we cannot hybridise wheat and coffee. Wheat can also be immune, for example, to a parasite of rye and this immunity can be diluted by hybridisation of wheat and rye; but immunity is an absolute quality; logically, this second category of apparent, dilutable immunity cannot be called immunity. The term "polyphyletic resistance" is proposed for its description.

Just as an non-host is immune, so a non-parasite is non-parasitic. Non-parasitism, like immunity, is an absolute quality. When it is dilutable by parasite hybridisation, it may be termed polyphyletic parasitic ability (or pathogenicity).

7.3 Examples

If we wish to study polyphyletic pathosystems, we should look for old records of "physiologic specialisation" which is not due to a vertical pathosystem and which occurs in crops of polyphyletic origin.

7.3.1 Sugarcane

When Abbott (1938) was studying red rot of sugarcane in Louisiana, he commented that "study... showed that the failure of POJ 213 was associated with specialised physiologic races and led to the recognition of the fact that, because of the great variability of the causal fungus, resistance to red rot might not be depended upon as a permanent attribute of any variety". In other words, he was prophesying a boom-and-bust cycle of cane cultivar production which, in nearly forty subsequent years has not occurred. His published data also indicates that the failure of POJ 213 was due to an initial disease escape. This was the result of inadequate testing of a foreign cultivar, and the accidental issue of disease-free seed setts of a cultivar with high leaf resistance but low stalk resistance.

Nevertheless, Abbott (1938) made many hundreds of pathogen isolates, undertook many inoculation experiments and did record physiologic specialisation. The isolates differ morphologically between the extremes of "dark form" and "light form". Cane cultivars have a clear polyphyletic origin. The early POJ varieties (Proefstation Oost Java) involved hybridisation with *S. robustum* and the early Co. varieties (Coimbatore, India) involved hybridisation with *S. spontaneum* and *S. barberi* which is believed to be a natural *S. officinarum* × *S. spontaneum* hybrid. There has also been much crossing of various cultivars and the entire genetic background of cane cultivars is confused. Unfortunately, the available data at the time of writing were inadequate for a clear demonstration of a polyphyletic pathosystem and further investigation is necessary. What is quite clear, however, is that Abbott's physiologic specialisation was not due to vertical resistance. The resistance is inherited polygenically; there is no gene-for-gene relationship and no Person differential interaction; and the resistance mechanisms are both complex and quantitative.

7.3.2 Wheat

The possibilities of polyphyletic pathosystems in wheat are interesting. Wheat itself is a polyphyletic hybrid produced during domestication. Some of its parasites, such as *P. graminis*, have many *formae speciales* identified by a differential interaction with related species of host. We should perhaps conclude that *P. graminis tritici* is itself a polyphyletic hybrid. These possibilities warrant investigation in view of the current interest in the wheat × rye hybrid (Triticale) which, theoretically at least, may be vulnerable to new polyphyletic pathotypes.

7.3.3 Coffee Berry Disease

We have seen that *C. arabica* is a simple allotetraploid derived from two wild, diploid progenitors. It was postulated that the extreme susceptibility of arabica coffee to the CBD form of *C. coffeanum* was due to a host erosion of horizontal resistance during at least fourteen centuries of cultivation in the absence of the parasite. The very high levels of disease suggest, however, that the CBD form is a polyphyletic pathotype derived by hybridisation between the two forms of the progenitor hosts, and possessing the natural maximum of horizontal pathogenicity. If this is not so, a future pathogen erosion of the resistance to CBD in Ethiopian coffee selections may be expected. Such doubts can only be resolved by appropriate interaction studies with the wild progenitors and, obviously, this is a matter of some practical importance.

7.3.4 *Tea*

Cultivated tea represents a simple quantitative polyphyletic origin. *Phomopsis theae* causes a collar and branch canker and it is most unlikely that vertical resistance would occur against a bark parasite of an evergreen, sub-tropical perennial. However, physiologic specialisation occurs (Punithalingam and Gibson, 1972) and different "biotypes" have various host ranges within tea clones. It is not yet clear, however, to what extent this may be due to a differential interaction between tea clones and environmental factors.

7.3.5 *The Fusarium Wilts*

F. oxysporum has one of the widest host ranges known to plant pathology and its hosts include Gymnosperms and, among Angiosperms, both monocotyledons and dicotyledons. *F. oxysporum* has been sub-divided into some sixty five *formae speciales* and has been reviewed by Booth (1971).

Economically, the most important monocotyledon hosts are banana (Panama disease), dates (Bayoud disease) and oil palm. The most important dicotyledon hosts are brassicas, flax, tomato, tobacco and cotton.

This pathogen is remarkable for the way in which the one term "physiologic specialisation" has been used to describe so many different kinds of differential interactions. The *formae speciales* are differentiated essentially on the basis of immunity. Each *forma specialis* is named after its principle host and may occasionally have a host range spanning several related genera; but, in general, the hosts of other *formae speciales* are immune to any one *forma specialis*.

At the other extreme, differential interactions due to a vertical pathosystem occur. Vertical resistance has been clearly established in cabbage, tomato, melon, watermelon, pea, soy bean and cotton. It may occur in other hosts also, although it is unlikely to be found in evergreen, tropical perennials with natural vegetative propagation, such as banana and dates.

Between these two extremes of immunity and vertical resistance, there are other kinds of differential interactions. In this respect, one of the most interesting Fusarium wilts is that of flax, which is caused by f.sp. *lini* and has been reviewed by Kommedahl *et al.* (1970).

In the United States, flax breeding for wilt resistance started in 1901 when Bolley observed that some flax plants grown on thoroughly infested soil survived when nearly all other plants died. The resistance of such plants was demonstrated and later developments included the establishment of pure lines and the release of resistant cultivars to farmers. However, resistant cultivars have not always remained resistant. Kommedahl *et al.* (1970) comment, quite correctly, that this can best be explained on the basis of "physiologic specialisation"; that is, a differential interaction. However, no one seems to have determined what kind of differential interaction is involved. It is not a vertical differential interaction; the inheritance of resistance is polygenic and the boom-and-bust cycle of cultivar production has ceased.

More data are necessary before firm conclusions can be drawn. However, it seems that this is a polyphyletic differential interaction which is complicated by other differential interactions due to environmental factors. It is possible that resistance mechanisms and polyphyletic pathotypes are both sensitive to environmental factors such as temperature. Resistance which is adequate in one season or one locality (i.e. one pathosystem) is then inadequate in another season or another locality.

Kommedahl *et al.* (1970) make two observations of particular interest. First, there has been a steady increase in resistance. Failures of resistance were common in the early cultivars and are rare or even non-existent in modern cultivars, from which we conclude that many resistance mechanisms were either lacking, or operating at a low level of effectiveness, in the early cultivars. These inadequacies presumably relate both to individual environmental factors and to individual polyphyletic pathotypes. In the course of continued breeding and selection, both the number of resistance mechanisms and their level of effectiveness increased, a phenomenon in accordance with the African maizes.

Secondly, seeds of ten morphologically distinct cultivars were mixed, sown in contaminated fields, harvested and planted again in the same or different fields for 4–5 years. After four years in the wilt nursery, six of the ten cultivars had disappeared and, of the remaining four cultivars, the two most resistant cultivars comprised 89% of all plants, a phenomenon also in accordance with the African maizes. Indeed, it is a most elegant demonstration of the effects of positive selection for resistance on a mixed host population.

Finally, we must compare flax breeding for wilt resistance with vertical resistance. There is no doubt that this breeding involved both horizontal resistance and a boom-and-bust cycle of cultivar production though it is, perhaps, an extreme exception. What is important is that the boom-and-bust cycle was not a true cycle; the sequence was asymptotic and both the frequency and the severity of the failures declined, modern flax cultivars now being fully resistant and the boom-and-bust sequence over. In terms of practical pathosystem management, it is important to be able to recognise this situation. Given clear recognition and suitable techniques, it will be possible to conclude the boom-and-bust sequence much more quickly than was possible with the inevitably blind, trial-and-error approach that was necessary with flax. Even more important, clear recognition will provide a degree of confidence in the selection work which was totally lacking for half a century of work on flax. In retrospect, the flax workers deserve the highest praise for perseverance and they have at last been suitably rewarded.

Chapter 8 Crop Vulnerability

Crop vulnerability means that the crop is unsafe in that there are one or more, more or less remote and currently inoperative pathosystem factors which threaten to unbalance the system, perhaps disastrously and, possibly, permanently. It is an essential aspect of pathosystem management to identify these threats, to assess them, and then to anticipate them in order to alleviate their eventual effects.

8.1 Description of Crop Vulnerability

8.1.1 Definition

We can define crop vulnerability as susceptibility in the absence of an epidemiologically competent parasite or pathotype; it is a concealed susceptibility.

Vulnerability may be natural or artificial. Vulnerability in a natural pathosystem occurs when the host and parasite have evolved separately, chestnut blight (*Endothia parasitica*) in North America being a typical example. More commonly, the concealed susceptibility is artificial and is due to the activities of man in the course of cultivating, domesticating or breeding a crop species in the absence of a parasite. Coffee berry disease in Ethiopia is typical.

8.1.2 Epidemiological Competence

The factor of epidemiological competence is crucial. High altitude and high latitude maizes are susceptible to *P. polysora* but they are not vulnerable to it because *P. polysora* has epidemiological competence only at low altitude and low latitude. The same is true of *P. graminis tritici*; the wheats of Northern Europe are susceptible but not vulnerable to this rust which lacks epidemiological competence at high latitudes, as well as at high altitudes in the tropics. Nor are the potatoes of northern Europe vulnerable to bacterial wilt (*P. solanacearum*), even though they are highly susceptible to this essentially tropical and sub-tropical pathogen. Similarly, it seems that the vertical resistance to stem rust in the spring wheats of north America is now frozen and that this parasite now lacks epidemiological competence on those cultivars in that area.

8.1.3 The Manifestation of Vulnerability

The susceptibility which is responsible for crop vulnerability is a hidden susceptibility which is revealed only when the absent parasite or pathotype is introduced to the pathosystem. Such susceptibility is a lack of pathosystem balance but this lack is concealed. When the parasite or pathotype appears, the susceptibility is revealed and the vulnerability is manifested.

The manifestation of vulnerability is normally an artificial event. The chestnut forests of North America were entirely safe until man interfered and actually imported *E. parasitica*; it could not get there any other way. Equally, the arabica coffee of Ethiopia remained free from CBD for some fourteen centuries. This pathogen could not cross the deserts of northern Kenya, northern Uganda or southern Sudan unless carried by man, as happened in the late 1960s, probably on coffee seed imported from elsewhere in Africa. As a general rule, it can be assumed that, if a parasite is capable of a natural dispersal to a new area, it will have reached that area by now.

The most important movement of parasites stems from ignorance, greatly enhanced in recent years by the development of air travel. In the industrial countries, the worst offenders are perhaps the amateur horticulturists who seem to have a morality all of their own. Taking cuttings from a neighbour's hedge is not really stealing; and taking cuttings through the customs in a sponge-bag is not really smuggling. In the developing world, the worst offenders, tragically, are the agricultural scientists themselves. They have been responsible for more plant movement than any other group and, far too often, the plant introductions have been careless. This category of parasite introduction is thus accidental, even if the accident is due to ignorance or negligence. Some parasite movement is deliberate, usually for purposes of research, but it must be noted that this kind of parasite introduction is generally carefully controlled and the risks are slight, although blue mould of tobacco *(Peronospora tabacina)* was introduced to Europe in this way. There is also the possibility of a malicious introduction and the less said about this the better. Finally, the possibility of biological warfare must be noted. During World War I, there was a proposal to drop Colorado beetles from aircraft on to the potato crops of Germany. Fortunately, civilised councils prevailed; but they may not always prevail.

8.2 Categories of Vulnerability

8.2.1 Absence of the Parasite

This is the most important category of crop vulnerability, as many major parasites have a restricted geographical distribution. The best summary of this situation will be found in the World distribution maps of parasites issued by the Commonwealth Agricultural Bureaux. Unfortunately, the preparation of even one map involves many, highly specialised manhours of work. Even though several hundred maps have now been published, the series is far from complete, which indicates the magnitude of the overall problem.

Important parasites may be geographically restricted for four different reasons. First, as we have seen, they may have evolved in one part of the world only, the classic example being chestnut blight.

A host may be moved and encounter the indigenous parasites of its relatives in a new area; Moko disease of banana has already been discussed in this context.

This second kind of geographical restriction has one great advantage, the danger is clear. Had the American chestnut *(Castanea dentata)* been taken to Eastern Asia, the dangers of *E. parasitica* might have become apparent. Because bananas were taken to Central America, all other banana-growing countries are now aware of the dangers of Moko disease.

Third, a host may be moved, leaving one or more of its parasites behind, only to constitute an increasingly dangerous threat. The vulnerability increases as the crop increases in size and value, often because of the parasite-free environment. The vulnerability also increases as the crop susceptibility increases, due to both host erosion of horizontal resistance and cultivation beyond its ecological limits. This category of geographical restriction of parasites is possibly the most common and the most dangerous. Coffee leaf rust *(H. vastatrix)*, coffee berry disease *(C. coffeanum)*, South American leaf blight of rubber *(M. ulei)* and blister blight of tea *(E. vexans)* are typical examples.

Lastly, certain components of one parasite species may be geographically restricted. Potato blight *(P. infestans)* now has a worldwide distribution but only as one mating group, the second mating group and, hence, functional oospores occurring only in Mexico. Were they to be introduced to Europe or North America, the initial inoculum would be so increased that "late blight" may become "early blight" and levels of horizontal resistance which are now adequate to control the disease may become inadequate. Functional oospores also increase the vertical mutability of the pathogen and controlled patterns of vertical resistance would become less valuable. We should note that, unless the second mating group lacks epidemiological competence outside Mexico, which is unlikely, *P. infestans* was exported from Mexico only once, as one mating group. The chance of more than one transfer, each time as one mating and the same mating group, is remote. It was exported to Europe, and it seems that all subsequent transfers derived from the one mating group in Europe, indicating that the artificial movement of parasites is perhaps less frequent than we think. Some areas of the world remained free from blight until very recently.

8.2.2 *Absence of a Vertical Pathotype*

If a parasite is present and the host crop is protected by a single vertical resistance combined with a low horizontal resistance, the crop is vulnerable to a matching vertical pathotype. This is the basis of the boom-and-bust cycle of cultivar production and is a key feature of pathosystem mismanagement. It must be appreciated that we are still in the era of the boom-and-bust cycle, and vast areas of wheat, rice and other crops are vulnerable in this way.

It is this form of crop vulnerability which is responsible for the myth of the dangers of crop uniformity. Given effective pathosystem management, a vertical pathodeme should have enough horizontal resistance to control the esodemic, so that the crop is no longer vulnerable and crop uniformity is no longer dangerous.

8.2.3 *Absence of a Horizontal Pathotype*

This category of vulnerability is unimportant and is mentioned only for comprehensiveness. If new cultivars were produced by screening with horizontal pathotypes of less than the maximum horizontal parasitic ability, there would be a subsequent parasite erosion of horizontal resistance. In general, this would happen quickly and the effects would consequently be slight because there would not have been enough time for the cultivar in question to become important.

8.2.4 Absence of a Polyphyletic Pathotype

The absence of a polyphyletic pathotype could lead to disappointment with new hybrids such as Triticale, and also to a declining boom-and-bust sequence of cultivar production, as apparently occurred with flax wilt. In general, this category of crop vulnerability is rare and is only occasionally likely to prove important.

8.2.5 Environment Erosion of Horizontal Resistance

Some crops can be cultivated well outside their ecological limits but only in the absence of a major parasite. Arabica coffee, for example, can be grown in the much wetter and warmer environment of robusta coffee, but only if rust *(H. vastatrix)* is absent. When rust is present, it seems that even the highest available levels of horizontal resistance are inadequate to protect arabica coffee growing in a robusta coffee climate. This is well-attested in Africa and especially in Ethiopia, which is the centre of diversification and where horizontal resistance to rust is at its natural level. The arabica coffee industries of Indonesia and the Philippines failed on the appearance of rust because they were located in robusta coffee climates. Much of the new world arabica coffee is vulnerable in this way, but new world coffee being cultivated in arabica coffee environments will probably survive the manifestation of its vulnerability to rust.

8.2.6 Abnormal Loss of Tolerance

The extreme sensitivity to rare mycotoxins in maize (Race T of *H. maydis*) and oats *(H. victoriae)* has already been discussed. In each case the loss of tolerance was linked to a single inheritance factor controlling a valuable character. Clearly, this is a special and very rare form of crop vulnerability. There must be many thousands of other crop characters whose inheritance is controlled by a single gene which does not have an abnormal loss of tolerance linked to it. These two exceptions prove the generality of this rule.

8.2.7 Apparent Resistance

Occasionally, a new cultivar is released to farmers in the genuine but mistaken belief that it is resistant. This happened in Louisiana with the sugarcane cultivar POJ. 213 which was believed to be resistant to red rot. Ironically, this unnecessary form of vulnerability is most likely to occur after a parasite has been successfully controlled with horizontal resistance. As we have seen, sugarcane diseases were controlled and then forgotten, but there have been several resurgences of sugarcane mosaic, for example, due to the release of mosaic-susceptible cultivars. The scientists responsible had possibly forgotten that mosaic even existed.

8.2.8 *The Common Factor*

It will be observed that, in all seven categories of crop vulnerability, there is one common factor which is an inadequacy of horizontal resistance. Given sufficient horizontal resistance, there would be no crop vulnerability.

8.3 Assessment of Crop Vulnerability

Crop vulnerability must be assessed if effective pathosystem management is to be achieved. When *P. polysora* reached Africa, there were fears of a famine comparable to the potato famine in Europe. These fears did not materialise because the disease was not understood; its environmental limits were not appreciated and the nature of the horizontal resistance was unknown.

There are two human factors which should be remembered in this context. The first is undue pessimism. Plant pathologists and entomologists tend, very naturally, to over-emphasise the importance of parasites and we all tend to remember disasters but to overlook non-disasters. It used to be a favourite pastime of tropical plant pathologists and entomologists to list new records in their annual reports. This was a period when many "new" crops were being tried in every developing country and many crop introductions resulted in "new" parasites. For every important parasite introduction, however, there were many unimportant introductions. Nevertheless, phytosanitary committees draw up long lists of alien parasites and the potential importance of many of them may be over-estimated.

The second human factor is undue optimism, which is most common among plant breeders who tend, very naturally, to relish the popularity of their latest cultivar and who hope, against all the evidence, that its vulnerability will not be manifested.

8.3.1 *Numbers of Alien Parasites*

All the alien parasites of a crop species must be enumerated and assessed. If the crop is indigenous, alien parasites will be few, but they are likely to be important because they have evolved separately from the host, as with Moko disease of banana. If the crop is exotic, the potential effects of alien parasites must be assessed from a study of their foreign pathosystems.

8.3.2 *Intensity of Damage*

The probable intensity of damage is assessed by a detailed consideration of the three main components of the pathosystem.

The most important factor in the host population is host erosion of horizontal resistance. If this has occurred in the absence of the parasite, the unselected level of horizontal resistance will normally be reached in 10–15 host generations. The same is true of the vertifolia effect in vertically resistant crops.

Environmental factors can also be assessed. It should have been possible to forecast the environmental limits of *P. polysora* in Africa from experience in the Americas, and it is now clear that *P. polysora* lacks epidemiological competence in much of Africa. Equally, it should have been possible to forecast the intensity of disease in areas where the pathogen was epidemilogically competent. Such hindsight is relevant when it comes to assessing the intensity of damage when vulnerability is not yet manifested.

The parasite itself is unlikely either to gain or lose parasitic ability in the course of being moved from one area to another and, apart from its existence, it is the least important variable to be considered when assessing the intensity of damage.

8.3.3 *Duration of Damage*

The duration of damage depends on the speed of recovery from the parasite. Clearly many social, economic and pathosystem factors contribute to this and a detailed enumeration of these is not necessary; occasionally, as we shall see in a moment, there is no recovery.

8.3.4 *Scale of Measurement*

It is clear from this discussion that the degree of crop vulnerability can range from the trivial to the disastrous. It would be useful to have a scale of crop vulnerabilities. based on the damage caused in the event of the vulnerability being manifested. The following scale of four categories is suggested:

Slight: the damage is financial only and can usually be absorbed by the growers; losses can generally be made up by improved productivity.

Moderate: the damage is still financial only but involves other economic sectors such as the processing and distribution industries; losses can generally be absorbed by the industries of the country concerned.

Severe: the damage is social as well as economic; social damage includes unemployment and malnutrition; economic damage includes significant loss of overseas trade, etc. In developing countries, this level of damage would necessitate additional overseas aid.

Disastrous: both the social and the economic damage is crippling. Social damage may involve mass poverty and famine, economic damage involves the ruin of a major industry, possibly the only significant industry.

8.3.5 *Type of Crop*

As a general but by no means invariable rule, the most severe and disastrous crop vulnerabilities occur in tree crops. There are three reasons for this. (1) The crop has probably developed major economic significance because of its freedom from parasites. (2) Breeding and selection has been conducted mainly for yield and quality in the absence of the parasite and host erosion of horizontal resistance has increased the vulnerability. (3) Tree crops are both the most valuable crops and the most difficult to replace. Recovery from the manifestation of crop vulnerability is thus difficult and may not occur.

Many of the industrial crops are approaching a marginal lack of competitiveness with synthetic substitutes. The manifestation of a crop vulnerability can then have three effects. First, the cost of artificial, parasite control methods, combined with a world shortage of the crop product, leads to price increases such that a previously uneconomic industrial substitute becomes economic. Manufacture of the substitute then increases greatly. The improved manufacturing techniques, resulting from this increase, may so reduce the costs and difficulties of manufacture, that the natural product is no longer competitive, even when the manifested vulnerability has been overcome with new, resistant cultivars. There is then no recovery from the manifestation of vulnerability.

8.3.6 Geographical Factors

Crop vulnerabilities are generally less along the same degree of longitude, where the climate varies from extreme polar cold to extreme tropical heat, than along the same degree of latitude, where the variation in climate and epidemiological competence is much less. Large oceans constitute an excellent barrier to parasite transfer and, for this reason, the dangers between continents are often greater than those within continents.

8.3.7 Type of Country

As a general rule also, crop vulnerability is greatest in tropical countries because these are developing countries. Circumstances of poverty, heavy reliance on agricultural exports, the all-too-frequent dependence on a one-crop economy and the difficulties of recovering from a manifestation vulnerability, all aggravate the problem.

8.4 Reduction of Crop Vulnerability

8.4.1 Phytosanitation

Phytosanitation is the control of the movement of plants with a view to preventing the spread of crop pests and pathogens. Phytosanitary control can operate at four levels.

1. International phytosanitary agreements involve all the countries within a well-defined phytosanitary area such as South America, Africa, etc. Such agreements are only as safe as the weakest member country. An island continent with only one government, such as Australia, can exert far more effective phytosanitary control than, say, Africa or South America.

2. National phytosanitation is primarily concerned with controlling imports of plant material, and is usually operated by the existing customs service often assisted by special plant inspectors, at all ports, airports and road entries. It can be highly effective but, inevitably, some failures occur. National phytosanitation can also provide varying degrees of phytosanitary certification for agricultural exports. These are usually dictated by the purchaser and, if the export is an important one, this control can also be highly effective.

3. Regional phytosanitation is designed to prevent plant movement from one region to another within one country. It is rarely effective because it is not feasible, for example, to set up road blocks and to search every car.

4. Phytosanitary control can be exercised at the farm level. This is valuable only for parasites with low dissemination efficiencies such as potato wart disease, and can be highly effective because one farm is usually controlled by one man. It can operate for both inward and outward movement and it may be governed by legislation.

Phytosanitary regulations recognise various categories of plant material, the most dangerous being material for propagation and this category has four subdivisions. The first and safest is true seed. Many parasites are not seed-borne and, of those which are, many can be destroyed by seed treatment. The second, and less safe, is unrooted cuttings; the third, and still less safe, includes various tubers, corms, rhizomes, etc., used for vegetative propagation. The last and most dangerous is plants actually growing in unsterilised soil. Other categories of plant material include materials for consumption, packing material, timber and so on, which are generally not dangerous.

Phytosanitary regulations recognise various categories of restriction on plant import, including unrestricted entry; import only with a special phytosanitary certificate; release only after inspection and treatment; plant quarantine only; and total prohibition. In their turn, these restrictions are affected by the country of origin and the parasites known to occur in it. Some confusion is caused by international trade in seeds in which an exporting country is not necessarily the country of origin.

It is clear that phytosanitation is a holding operation only; eventually, all parasites will penetrate to all parts of the world where they are epidemiologically competent. A holding operation is valuable nevertheless, even if it does no more than allow the time necessary to reduce crop vulnerability in other ways.

8.4.2 Quarantine

An effective plant quarantine station has four attributes. First, it must be safe; no parasite must be allowed to escape from it. This is best achieved by geographical isolation. For example, tropical plants can be quarantined in temperate countries and *vice versa*. Other precautions include specialised equipment such as spore-proof growth chambers.

The quarantine station must be a filter and not a barrier. Many of the older stations merely grew the imported material and, if it was in any way suspicious, it was destroyed. Such a station is a barrier. Repeated imports of valuable plant material may be destroyed and the station has stopped the import of a parasite but no more. A filter quarantine lets through the host and holds back the parasites, a process which may be expensive in both time and money as it often involves difficult techniques such as meristem culture. But it is clearly necessary, except for the most frivolous imports.

An effective quarantine is two-directional, guaranteeing plant material which is being exported as well as material being imported. There are obviously quantitative limits to such activity which should normally be confined to valuable breeder's material in which high crop vulnerability is involved.

Lastly, a plant quarantine station should be international in the widest and best sense of science being for the benefit of all mankind. Civilisation is one system.

It must be said that no such quarantine station exists although a few approach this ideal fairly closely. Ideally, there should be an international plant quarantine station for each major crop (or group of related crops) in which there are severe or disastrous phytosanitary risks in the world. The station should be located in a country where the crop concerned is of no economic importance.

8.4.3 Preparation

Crop vulnerability can be reduced by various methods which anticipate the parasite introduction and prepare for it.

The most important preparation is to avoid the host erosion of horizontal resistance which occurs in the absence of the parasite. This is achieved by breeding, but it is difficult because all resistance screening must be conducted in another country where the parasite is present. Thus, new cane cultivars produced in Hawaii are sent to Fiji for assessment of their susceptibility to Fiji disease, so that in the event of this virus appearing in Hawaii, there will be advance information on susceptibilities and the best replacement cultivars. When the crop is a valuable perennial, such as Hevea rubber in Malaysia, and highly vulnerable to an alien pathogen such as South American leaf blight *(M. ulei)*, it is necessary both to produce new, resistant, high-yielding cultivars suitable for the local environment, and to use them as widely as possible in all new planting and normal replacement planting. Ferwerda (1969) has described the elaborate process of breeding rubber in Liberia, Ceylon and Malaysia, testing progenies for resistance in the Americas with intermediate quarantines, etc. The cost of replanting all rubber plantations after the disease has appeared may well be prohibitive, but if a significant proportion of the plantations consist of resistant host material, the crop vulnerability will be reduced. Comparable preparation work should be undertaken with tea in Africa against blister blight, coffee in South America against coffee berry disease, and various other parasites which are responsible for severe or disastrous crop vulnerabilities.

Preparation includes the possibilities of parasite eradication after its introduction. Hutchinson (1958) has described a plan for defoliating rubber trees in the event of *M. ulei* appearing in Malaysia, with a view to eradicating the pathogen. In general, however, the possibility of successful eradication of a polycyclic disease of high dissemination efficiency is remote. When the parasite dissemination efficiency is low, as with coffee berry disease, or a soil-borne parasite, the chances of success are greater. Both early detection and ruthless destruction are necessary if any attempt at eradication is to be successful. The successful eradication of coffee leaf rust after its appearance in Papua New Guinea (Shaw, 1973) should be noted.

Finally, the experience with Race T of *H. maydis* in the USA emphasises two essential points. Resistant material must be available and the seed multiplication and distribution infrastructure must be effective if the results of manifestation of crop vulnerability are to be overcome. This is particularly important with the vulnerability due to the breakdown of vertical resistance in annual crops.

Chapter 9 Conclusions

9.1 The Value of the Systems Concept

9.1.1 Properties of Patterns

The value of the systems concept is that our comprehension of complex systems is enhanced if we analyse them in terms of the simple properties of patterns. It will now become apparent that there are three aspects of systems analysis which are of particular importance to our understanding of pathosystems. These are systems levels, systems balance and systems control.

9.1.2 The Importance of Systems Level

Depending on the basis of the systems analysis, it is possible to define various series of systems levels in a pathosystem. These series usually correspond more or less closely with each other.

The first series of levels concerns the host. The highest systems level is the gene pool. Next is the population; then the individual host plant; the individual cell; and, finally, the biochemical components of that cell.

The second series concerns "fitness" in the Darwinian sense of the word. Among cultivars, fitness is usually called domestication and, obviously, it is only the fittest (or best domesticated) which survive. At the highest systems level, domestication is a single survival value; some cultivars are better than others. At the next level, domestication has the three components of yield, quality and resistance. At the next, lower level, each of these components has many subcomponents. Resistance, for example, is made up of many different resistances to many different parasite species. At a still lower level, each resistance to one parasite species is due to many different resistance mechanisms and, still lower, each mechanism has many biochemical components.

The third series of systems levels concerns the breeding technique. The highest is represented by the maizes of Africa in which the accumulation of horizontal resistance was so close to the natural pathosystem that no scientists were involved. The next systems level is represented by the coffee programme in Ethiopia in which scientists were involved but only in the selection, testing and propagation of desirable individuals within an existing population. The next level is represented by population breeding in which one population is randomly polycrossed with another. Population breeders are primarily concerned with biometrics and polygenically inherited characters, and this systems level involves a further increase in the amounts of scientific labour and expertise. The next level is pedigree breeding in which one individual plant is crossed with another and Mendelism and oligogenically inherited characters are favoured. This level is even more labour-intensive and skilled. A still lower level involves the individual resistance mechanism and it is at this level that the histologists begin to cooperate with the breeders. They identify prominent mechanisms and devise techniques for their recognition and measurement so that the breeders can screen progenies on the basis of their presence or absence. It is at this level that the breeders become most interested in the genetics of the parasite, the gene-for-gene relationship and vertical resistance. Breeding at this level is extremely labour-intensive. Finally, at the lowest systems level, the biochemists enter the picture and investigate the chemical pathways of one mechanism. At this level, the work is so complicated and involves such sophisticated techniques that contributions to the prevention of crop loss have still to be made.

9.1.3 *The Importance of Systems Balance*

Systems balance is clearly essential at all systems levels. At the highest level, only the best cultivar is cultivated and no farmer can be expected to cultivate an inferior cultivar. At the next level, this means that the best cultivar has a high yield, quality and resistance and, if any of these components is deficient, the cultivar is defective. At a lower level, high resistance means resistance to all locally important parasites and, if only one resistance is deficient, the cultivar is susceptible. As we saw earlier, it really matters very little which parasite is causing the destruction; it is the destruction itself which is important. At a still lower level, the resistance to any one parasite can only be maximal if all mechanisms are present and operating at their highest level of efficiency.

9.1.4 *The Importance of Systems Control*

Systems balance is maintained by systems control and the balance can be lost if the control is inadequate or inappropriate.

We saw earlier that there are two basic kinds of pathosystem. The natural or wild pathosystem (disease triangle) is characterised by its good systems balance because, were it not balanced, it could not have survived evolutionary competition. The artificial or crop pathosystem (disease square) is conspicuous by its lack of systems balance and cultivars are notoriously less resistant than wild plants.

The fundamental difference between these two kinds of pathosystem is in their systems control. The natural pathosystem is entirely autonomous and the crop pathosystem is the result of at least some deterministic control. It seems also that, as we descend from the highest to the lowest systems level in our breeding, the amount of deterministic control increases and the loss of systems balance also increases. At the one extreme, there were no scientists at all controlling the maizes of Africa and a balanced pathosystem resulted while, at the other extreme, breeding for a single, prominent resistance mechanism, inherited by a single gene, leads to vertical resistance and the boom-and-bust cycle.

If we accept that the lower the systems level, the greater is the deterministic control and the greater the loss of systems balance, we must conclude that "science" is inimical to systems balance and that the more fundamental that science, the more inimical it becomes.

The crucial factor is the complexity of the system. Deterministic control can lead to perfect systems balance in a simple system, such as the aerodynamic stability of an aircraft, or the constant voltage of an electricity supply. But, in a complex system, deterministic control is liable to be exerted unevenly, with some components being over-controlled and others under-controlled. There is also a close quantitative relationship between the loss of balance and the systems level at which the deterministic control is exerted. At the highest level, domestication is only one survival value. At the next level, it has three survival values; yield, quality and resistance. At the next level, there are many different resistances; still lower, there are many different mechanisms to each resistance. The lower the systems level, the more components there are to be controlled deterministically and the less is the possibility of achieving a balanced control.

However, behind and obscured by our uneven and clumsy attempts at deterministic control, every biological system has an autonomous control which, if left alone, will achieve a perfect systems balance. There seems to be an inescapable conclusion; if we wish to achieve pathosystem balance, we must work at the highest systems level and we must reduce deterministic control to the minimum.

9.1.5 The Evidence Reviewed

The highest systems level is represented by the maizes of Africa in which an entire gene pool changed from high susceptibility to high resistance to *P. polysora* during some 10–15 generations. Mortifying though it may be, this happened in the absence of breeders and pathologists; the deterministic control was negligible and a perfect pathosystem balance resulted. We recognise, however, that these maizes are subsistence cultivars and that further improvements in domestication are both possible and desirable. Given a minimum of deterministic control, or artificial selection, yields and quality could both be increased; so too could the levels of resistance which tend to the natural level of horizontal resistance. The exertion of the necessary selection pressures is so simple and "unscientific" that it could be done by the peasant farmers themselves, merely by detasselling the least desirable majority of the population and the selective harvesting for seed of the most desirable minority in each successive crop.

The next systems level is represented by the coffees of Ethiopia in which a large and genetically variable population had been cultivated for some fourteen centuries with negligible parasite control measures. A good pathosystem balance existed but it was upset by the introduction of coffee berry disease. The amount of deterministic control in selecting, testing and propagating desirable individuals within an existing population was appreciably greater than in the African maizes. Its main effect was to reduce time scales; if left alone, the Ethiopian coffees would have behaved like the maizes of Africa but, being a perennial crop, this process would have required a century or more and, in the meanwhile, the farmers were faced with ruin. As with the maizes, this deterministic control will also lead to improved yields and quality; an improved domestication.

The next systems level is population breeding in which there is considerably more deterministic control. Population breeders are inclined to deny the possibility of a balanced domestication; they either argue that comprehensive horizontal resistance is impossible to achieve or that, if it is achieved, this can only be done at the expense of yield or quality or both. Hawaiian sugarcane breeding is the exception which proves this rule of general pessimism.

At the systems level of pedigree breeding, the amount of deterministic control is approaching its maximum. The loss of systems balance is extreme, at all systems levels; between the three components of domestication, between resistances to different parasites, between vertical and horizontal resistance, and between resistance mechanisms. This distortion of systems balance has led to a corresponding distortion of scientific opinion, in which it is all too commonly accepted that any permanence in resistance breeding is impossible; that comprehensive resistance to all locally important parasites is also impossible; and that the possibility of resistance which is both permanent and comprehensive is beyond serious consideration.

Finally, at the lowest systems level, the biochemical investigation of individual resistance mechanisms is so divorced from practical pathosystem management that it is unlikely ever to contribute to the prevention of crop loss. We have seen that every African maize plant which was highly susceptible to *P. polysora* was highly resistant to *P. sorghi*. We have no idea of how many mechanisms were involved; we have even less idea of the nature of these mechanisms and still less do we understand their biochemistry. Nor does there appear to be any valid reason why we should.

It is quite commonly postulated that the best science is the most fundamental science and, in this context, fundamental means the lowest systems level. This is an issue which every scientist must decide for himself. It is the contention of this book, however, that the best science is also the most useful science and that, when it comes to pathosystem management, our concern is to feed a hungry world.

There seems to be little doubt that, as scientists, we have been altogether too clever and that, if our error is to be reduced to one word, the most appropriate is "specialisation". Our over-specialisation has led to fragmentation and our management of the one system has quite literally gone to pieces. We have allowed ourselves to become obsessed by details as we have delved deeper and deeper to the lowest systems levels, and the resulting complexity has proved entirely unmanageable.

The remedy is in the holistic approach, which is the antithesis of specialisation. If the remedy is to be reduced to one word, it is, perhaps, "simplicity". We must work at the highest systems level and exert only the minimum deterministic control necessary to achieve a balanced domestication. Good pathosystem management is so simple, and its benefits loom so large, that we have been unable to see them at all. It seems that even scientists can be blinded by science.

9.2 The Domestication of Resistance

9.2.1 Balanced Domestication

We have seen that the process of domestication involves a change of systems balance. Certain variable survival values, which contribute to competitive ability in a wild ecosystem, are reduced; and other survival values, which contribute to agricultural yield, quality and resistance, are increased. The ultimate control is the same in each system; it is due to competition and the survival of the fittest. However, the definition of fitness varies with each system. It follows that systems balance must be defined in terms of the requirements of that system.

A cultivar cannot survive in a wild ecosystem because it lacks systems balance outside the agricultural system. At a lower systems level, the requirements of one ecosystem are different from those of another; and, at a still lower level, the requirements of one pathosystem are different from those of another. When we refer to systems balance, therefore, we are referring also to a particular system and the particular requirements of that one system. Differences of pathosystem requirement have already been mentioned and are due to different distributions of parasites. Similarly, crop systems have differing requirements; in specialised horticultural crops, quality is often more important than yield or resistance; in extensive agriculture, yield tends to be the most important requirement; and, in subsistence crops, a high susceptibility to only one parasite means that the crop is not grown at all.

In spite of these qualifications, we can define a balanced domestication as one in which the three variable components (yield, quality and resistance) have been raised above the natural levels in the wild progenitors, according to requirement. This is done at the expense of survival values which are redundant in cultivation, and which are identified by the artificial components of cultivation, such as ploughing, sowing, fertilising, weeding, harvesting and seed storage.

9.2.2 Comprehensive Horizontal Resistance

Comprehensive horizontal resistance represents a balanced horizontal pathosystem, at all systems levels. At the highest level, the resistance is as good as the yield and quality; at a lower level, all the resistances to all the locally important parasites are equally high; and the same is true of all the mechanisms of any one resistance.

If we regard comprehensive horizontal resistance as a single survival value, it is one which is variable and which can be domesticated. By domestication, we mean that it can be increased, without loss of balance, above its natural level, which is enough (but only enough) resistance to ensure the evolutionary survival of the host in a natural pathosystem. In the crop pathosystem, we need higher levels of comprehensive horizontal resistance for two reasons. Many factors in cultivation, such as monoculture, crop uniformity and high host population densities, tend to increase the epidemic. Secondly, the evolutionary survival of the host is no longer a significant criterion; the only important criterion is that neither the yield nor the quality of the crop product are damaged by parasites. In practice, this requirement is likely to demand a crop husbandry which is virtually parasite-free, and this, in its turn, means that the domesticated levels of horizontal resistance must be considerably above the natural levels.

It is clear that some crop pathosystems are easier to manage than others, and it is now obvious that adequate levels of comprehensive horizontal resistance can be achieved, at least in some of the more manageable pathosystems. The benefits of such an achievement should be briefly noted.

The most immediate effect of comprehensive horizontal resistance is the virtual and permanent elimination of the crop loss due to parasites, as well as the costs of artificial control measures, at both the farmer and the national levels. Secondary benefits then become apparent. The toxic hazards, environmental pollution and pesticide-resistant strains of parasites, which result from the use of pesticide chemicals, are also eliminated. The security of production is increased by a reduction in seasonal fluctuations in yield. Increased monoculture and increased uniformity of crop and crop product become possible without crop vulnerability. Further increases in yield become possible by the use of increased inputs, such as sowing rates, fertilisers and longer growing seasons. Soil structure can be improved and soil erosion reduced by the incorporation of crop residues which are currently burned as a parasite control measure. Finally, the negligible cost of comprehensive horizontal resistance, whether in cash expenditure, effort or technical expertise, is of great benefit to the farmer, particularly in the developing countries of the world. He may even be able to grow crops which he could not grow previously because of parasite problems which were beyond his ability to control.

9.2.3 The Reinforcing of Horizontal Resistance

In some crop pathosystems, an adequate comprehensive horizontal resistance may be difficult to achieve and the horizontal resistance must then be reinforced. In one sense, we are doing this already, with virtually all our crops, but we are doing it by default. Ultimately, the reinforcement of horizontal resistance is justified only when we have accumulated the maximum levels of it and have shown them to be inadequate.

The first choice in reinforcement is clearly vertical resistance provided that its effects are permanent and its use does not lead to crop vulnerability. Permanence is achieved by the use of strong vertical genes, in the form of spatial or sequential patterns. It may also be possible to achieve a frozen vertical resistance but this will be possible in relatively few areas and against relatively few parasites. The crop vulnerability which is caused by vertical resistance is due to the vertifolia effect and this can be avoided by screening in both the esodemic and the exodemic. That is, a vertical gene is transferred to a good cultivar by back-crossing and, in each generation, every individual must first be inoculated with a non-matching vertical pathotype (exodemic) to demonstrate the presence or absence of that gene. Those individuals which are vertically resistant must then be inoculated with the matching vertical pathotype (esodemic) to measure their level of horizontal resistance, and only the most resistant are retained.

If vertical resistance or, more specifically, strong vertical genes are not available, the horizontal resistance must be reinforced with artificial control measures. Only three points need be made. Firstly, some artificial control measures, such as seed dressings, are cheap, simple and effective while others, such as the repeated application of a foliar fungicide, are expensive and clumsy. It is the latter category which indicates the greatest need for resistance. Secondly, many artificial control chemicals confer a vertical protection, and if a horizontal protection is unavailable or otherwise undesirable, the vertical system must be properly managed. The strength of the vertical protection should be known and the possibility of spatial or sequential patterns of different strong protections should be investigated. Thirdly, the use of artificial control measures should be such that a complete and balanced pathosystem management is achieved; this is often called "integrated control".

Finally, biological control, in the narrow sense of the term, should be mentioned. Occasionally, a parasite is moved to a new area and proves exceptionally destructive because some of its hyperparasites were left behind. Biological control involves the identification, selection and cautious introduction of these hyperparasites with a view to restoring pathosystem balance.

9.2.4 Conflicts between the Components of Domestication

It is quite obvious that there is an absolute maximum of crop production, beyond which no further improvements are possible. As one limiting factor is eliminated, so another becomes prominent. This is the law of diminishing returns and it is probable that the final limiting factor, which is beyond our control, is the amount of solar energy received by the crop.

Before this limit is reached, however, it is highly probable that conflicts between the components of domestication will become apparent. It may become impossible to increase yield except at the expense of, say, resistance, and a few deliberate susceptibilities may then be justified, provided that they involve parasites whose artificial control is both cheap and simple. However, such a procedure can be justified only after it has been clearly demonstrated that further yield increases are impossible in any other way. Nor should this general argument be used to justify the lack of pathosystems balance in present-day cultivars which are still a long way from the ultimate potential of crop production.

9.3 Cumulative Crop Improvement

9.3.1 Repetitive Plant Breeding

The boom-and-bust cycle means that conventional plant breeding for vertical resistance is an essentially repetitive process. With such repetitive plant breeding, improvements in both the level and the balance of domestication are not impossible but are nevertheless very difficult to achieve and even more difficult to maintain. This is indicated by the very disappointing results obtained from some seventy years of such breeding in many important crops.

9.3.2 Cumulative Plant Breeding

Once a complete and balanced pathosystem management has been achieved by comprehensive horizontal resistance, either alone or suitably reinforced, plant breeding becomes cumulative and progressive. A good cultivar need never be replaced, except with a better cultivar. Provided that pathosystems balance is retained, this cumulative crop improvement can continue, no doubt with diminishing returns, until the ultimate potential of crop production is approached. This is the real significance of accurate pathosystem analysis and sound pathosystem management.

Chapter 10 Terminology

This book began with the description of a word as a basic pattern which has the two properties of spelling and meaning. It is appropriate to end on the same topic. Let us examine this basic pattern with a view to discovering what we mean by precision.

Most technical books end with an index in which key words are listed alphabetically, according to their spelling. With computerised information retrieval, it is now possible to index key words according to their meaning, using an appropriate computer language. This is a great advance. However, if one spelling has several meanings, the word is imprecise and such indexing becomes difficult, if not impossible. Our best test of precision, therefore, is the ease with which a term can be indexed according to its meaning.

Precision can and should occur at higher systems levels. Thus, by using a particular pattern of words, we can increase the precision of a multi-word term. Compare "Person/Habgood differential interaction" with "physiologic specialisation".

At a still higher systems level, we enter the realms of logic where precision is crucial. It is unlikely that logic will ever have a mathematical precision because words are unlikely ever to have the precision of numbers. Nevertheless, it is remarkable how the mathematical aspects of the biological sciences have developed in recent years while logic seems to have been taken so much for granted that it has been virtually ignored. The many examples of false syllogisms, already quoted, illustrate this point. Good science is impossible without sound logic, and sound logic is impossible without precise terms.

For this reason this book ends with a glossary, where some 300 terms are defined. They are all terms used in this book and they are defined for the purposes of this book.

Abnormal loss of tolerance	The extreme loss of tolerance associated with Race T of *Helminthosporium maydis* in maize and *Helminthosporium victoriae* in oats. See p. 76 for full discussion.
Absolute horizontal resistance	A level of horizontal resistance so high that there is an apparent immunity. Absolute horizontal resistance differs from immunity in that it is an extreme of a variable survival value. It can thus be reduced due to host erosion.
Absolute minimum horizontal resistance	A point on a scale measurement of horizontal resistance; it can probably be determined experimentally by exerting selection pressure for susceptibility on a genetically flexible host population.
Absolute susceptibility	A theoretical point on a scale of measurement of horizontal resistance; it is a complete absence of horizontal resistance such that the pathogen can grow through living host tissue without any hindrance whatever.

Accumulation of horizontal resistance	The converse of erosion. See also: Restoration.
Active resistance mechanism	A resistance mechanism which operates in response to infection and/or colonisation (e.g. a hypersensitive reaction). It is probable that all vertical resistance is conferred by active mechanisms but that not all active mechanisms confer vertical resistance. See also: Passive.
Adult plant resistance	Resistance which is manifested mainly in older plants and is less apparent in seedlings. All adult plant resistance confers horizontal resistance but not all horizontal resistance is due to adult plant resistance.
Aestivation	Survival during a summer or tropical dry season.
Aggressiveness	This term normally means horizontal pathogenicity or parasitic ability but has also been used to mean vertical pathogenicity or parasitic ability.
Alien parasite	A parasite which is absent from a pathosystem. See also: Crop vulnerability.
Allo-infection	Infection in which the donor (or infector) host is a different individual from the recipient (or infected) host individual.
Allo-polyploid	A polyploid derived from genetically dissimilar diploid parents.
Apparent erosion of horizontal resistance	An increase in disease or damage levels due to changes in agricultural practice, such as a reduction in rotation or burning of crop residues; an erosion of horizontal resistance may then be falsely postulated.
Apparent horizontal resistance	Vertical resistance which has an appearance of horizontal resistance. This includes the effects of quantitative vertical resistance; multilines; the reduced horizontal pathogenicity of a complex vertical pathotype; and frozen vertical resistance.
Apparent vertical resistance	Horizontal resistance which has an appearance of vertical resistance. This usually occurs when there is a differential interaction which is not a Person differential interaction, when there is a temporary, physiological failure of horizontal resistance, with a polyphyletic pathosystem, when inadequate testing leads to an apparent high resistance which proves illusory only later, or with mis-labelling or mixing of cultivars.
Artificial	Man-made; the converse of natural. The effects of a deterministic control super-imposed on a natural, autonomous system. Thus, artificial selection, artificial levels of disease and horizontal resistance, artificial pathosystem, etc.
Artificial classification	The deme-system and type-system are artificial, temporary classifications of populations c.f. the natural and permanent classification of taxonomists.
Artificial control of parasites	Any control method which is man-made, e.g. the use of protective chemicals.
Artificial selection	The results of man-made selection pressures exerted by screening.
Auto-infection	Infection in which the donor (infector) host individual is the same as the recipient (infected) host individual.

Autonomous control/system	Systems control which is entirely within the system itself and which is consequently an autonomous or independent system.
Autopolyploid	A polyploid derived from genetically similar or indentical diploid parents.
Balance	Dynamic equilibrium. A balanced system is stable; a loss of systems balance leads to instability and, in extreme cases, the system can become self-destructive.
Behaviour	Systems behaviour corresponds to the meaning of a word, as opposed to systems structure which corresponds to the spelling of a word.
Biotype	A subdivision of a species which is normally distinguished by criteria other than those of morphology. These criteria include those of physiology, parasitic ability, resistance, etc. Use of this term usually inplies that a precise definition is not possible.
Black nomenclature	The nomenclature devised by Black *et al.* (1953) for the naming of a series of vertical genes. Each gene is given a numerical value in chronological order on discovery; thus, 1, 2, 3, etc. The name of each vertical genome is normally enclosed in brackets e.g. (1234) and, when more than nine genes are known, the numerals are separated by commas e.g. (1, 2, 12). Abbreviation is possible by representing all intermediate numerals of a continuous series with a hyphen e.g. (1–5, 7–11). Matching vertical pathotypes and pathodemes thus have the same name which also indicates the composition of their vertical genomes.
Boom-and-bust cycle	The cycle of repetitive plant breeding in which a new cultivar is produced to replace one whose vertical resistance has broken down.
Boom-and-bust sequence	The asymptotic rate of cultivar failures due to polyphyletic vulnerability.
Boundary	A system's boundaries are its operating limits; boundaries may be conceptual, geographical, physical, climatic, biological, etc. Thus, an ecosystem can be conceptually limited to macroscopic or microscopic phenomena, within a defined area, which either includes or excludes edaphic factors, during one or more seasons, in the presence or absence of stated parasites, and so on. It has been suggested, for example, that immunity is outside the conceptual boundaries of a pathosystem.
Breakdown	Some resistance mechanisms (which might confer either vertical or horizontal resistance) fail to operate under extreme environmental conditions. This phenomenon must be clearly distinguished from a breakdown of vertical resistance.
Breakdown of vertical resistance	A failure in the effectiveness of vertical resistance due to allo-infection with a matching vertical pathotype. The esodemic then begins. The term is a colloquialism and this may be indicated by single quotation marks. Strictly, it is the effectiveness of the resistance which breaks down, not the resistance itself. See also: Recovery.

Breakthrough	A Hegelian change in which a small qualitative change leads to a qualitative change. The term can be used in military, economic, scientific and other contexts.
Buffered system	If a dynamic system is able to exhibit wide swings away from its optimum, stable state and then recover from those swings, it is described as being buffered.
CBD	Coffee berry disease.
Chronological nomenclature	A nomenclature of vertical genes in which the genes are named in their order of discovery.
Closed system	In thermodynamics, a system which is unable to exchange heat across the boundaries with its external environment.
Colonisation	The invasion of host tissues by a pathogen following the successful completion of infection.
Competition	The converse of co-operation. Competition means that two systems co-exist in spite of each other; co-operation means that they co-exist because of each other. In a balanced, dynamic system, competition and co-operation between all sub-systems are normally equal.
Complete pathosystem management	Pathosystem management which ensures that the crop loss due to any one parasite species is reduced to the minimum and that this is true of all locally important parasites.
Complete vertical resistance	Vertical resistance which confers complete protection against non-matching vertical pathotypes. See also: Incomplete.
Complex vertical genome	Complex means that there are many vertical genes in the one genome; hence, complex vertical pathotype, pathodeme, pathogenicity (or parasitic ability), resistance, etc.
Composite cross	A plant breeding system in which many inbreeders are crossed in all combinations and are subsequently grown as a mixed, segregating population for a number of generations.
Comprehensive horizontal resistance	Adequate levels of horizontal resistance to all locally important parasites and pathotypes, including those which are absent, epidemiologically competent and the cause of crop vulnerability. However, the term does not include resistance to parasites which lack epidemiological competence in the locality concerned.
Constant ranking	The absence of a differential interaction and the definitive characteristic of Horizontal.
Constant sum of survival values	See: Hypothesis of
Continuity of host tissue	This term means that a host population has genetic uniformity in both space and time; vertical resistance has no survival value and, hence, cannot evolve in a host population which has both spatial and sequential continuity of host tissue. See also: Discontinuity.
Control	The effective communication with or between the components of a system such that systems balance is maintained. Control may be Autonomous or Deterministic.

Co-operation — The converse of competition. Co-operation means that two systems co-exist because of each other; competition means that they co-exist in spite of each other. In a balanced, dynamic system, co-operation and competition between all sub-systems are normally equal over a given period of time.

Critical state — The state of a system at which a Hegelian change is imminent. See also: Point.

Crop Esodemic — The esodemic at the pathosystem level of the crop in which all individual plants are the epidemiological equivalent of the individual leaves of one tree. The crop esodemic involves auto-infection within that one crop.

Crop exodemic — The exodemic at the pathosystem level of the crop in which each individual crop is the epidemiological equivalent of one host individual. The crop exodemic involves allo-infection between crops.

Crop loss assessment — A new discipline which is emerging within plant parasitology and which concerns the theory and practice of assessing the loss of crop from its parasites by examination of parasite population increase and damage.

Crop pathosystem — A pathosystem which is artificial due to deterministic control; cultivars differ from wild plants, cultivation differs from a natural ecosystem, and the parasite populations differ accordingly and may be artificially controlled. See also: Natural pathosystem.

Crop pattern of vertical resistances — A pattern of different vertical resistances in which each unit of the pattern is an individual crop. The pattern may be spatial, sequential or both. See also: Plant pattern.

Crop uniformity — Genetical continuity of host tissue in time, space or both. Crop uniformity reduces the value of vertical resistance which depends on both sequential and spatial discontinuity of host tissue. However, the term "crop uniformity" may legitimately refer to only some aspects of the crop (e.g. yield, quality) in spite of a non-uniformity in other respects (e.g. a multiline of several vertical pathodemes).

Crop vulnerability — Susceptibility in the absence of an epidemiologically competent parasite or pathotype. Crop uniformity increases but does not cause crop vulnerability.

Cultivar — A botanical variety of a cultivated, domesticated crop species. A cultivar is characterised by abnormally high levels in natural survival values contributing to the quantity and quality of the harvestable product. These high levels were gained by artificial selection at the expense of natural survival values contributing to competitive ability in a natural ecosystem.

Cultivation — The creation of an artificial ecosystem in which plants required by man are protected from much natural competition. See also: Domestication.

Cumulative plant breeding — The plant breeding which is possible with horizontal resistance and with which a good cultivar is replaced only with a better cultivar. See also: Repetitive plant breeding.

Cybernetics — The study of messages and, in particular, the effects of messages in systems control.

Cycle — The regular repetition of a phenomenon or process usually in the mathematical form of a harmonic motion from minimum to maximum to minimum; thus Disease cycle, Epidemic cycle, Boom-and-bust cycle.

Deme-System — A special purpose, artificial classification devised by Gilmour and Heslop-Harrison (1954); "-deme" denotes a population of individuals within a specified taxon. It must be used with a prefix which describes the criterion used to define that population; thus gamodeme, topodeme, etc. Robinson (1969) proposed that the deme-system should be reserved for the host and that a parallel type-system should be used for the parasite; hence pathodeme, pathotype.

Dependent system — A system which is controlled, at least in part, by external factors. Thus biological systems are dependent on solar system cycles.

Deterministic control — Systems control which is imposed by man with a view to achieving predetermined objectives. Deterministic control is a feature of all artificial (man-made) systems and is totally absent from all natural systems. See also: Teleology.

Differential hosts — A series of vertical pathodemes used for the identification of vertical pathotypes. Ideally, each differential host should possess only one vertical gene. The term can also be used to describe polyphyletic pathodemes and, in its widest sense, various host species which show a differential interaction based on immunity to various parasite species. See also: *Forma Specialis*

Differential interaction — When a series of different pathodemes is inoculated with a series of different pathotypes, there is a differential interaction if more than one pathotype is necessary to differentiate the pathodemes, and *vice versa*. See also: Person differential interaction, *Forma specialis*, Polyphyletic, Constant Ranking.

Differential interaction fallacy — The fallacy which arises from the fact that all vertical pathosystems show a differential interaction but that not all differential interactions are due to vertical resistance. See also: *Forma specialis*, Polyphyletic pathosystem, Person differential interaction.

Differential pathotypes — A series of vertical pathotypes used for the identification of vertical pathodemes. Ideally, each differential pathotype should lack only one vertical gene. The term can also be used to describe polyphyletic pathotypes and *Formae speciales*.

Discontinuity of host tissue — This term means that a host population has many vertical resistances in both space and time; vertical resistance has survival value only in a host population which has both spatial and sequential discontinuity of host tissue. See also: Continuity.

Disease — The adverse effects on a host due to a pathogen (which lacks mouth parts) as opposed to damage which is due to a pest (which possesses mouth parts) See also: Disorder.

Disease cycle	The cycle which starts with infection of the host and ends with reproduction of the pathogen. See also: Monocyclic, Oligocyclic and Polycyclic disease.
Disease escape	A relative lack of disease in a susceptible host due to an early maturation, the Rossetto hypothesis and other factors which prevent or reduce disease development.
Disease square	The artificial, crop pathosystem in which there are four primary elements of systems control: the host, the parasite, the environment and man.
Disease triangle	The natural, wild pathosystem in which there are only three primary elements of systems control; the host, the parasite and the environment.
Disorder	A non-parasitic, physiologic malfunctioning due to either an excess or a deficiency in an environmental factor such as wind, heat, light, water or nutrients. Disorders are outside the conceptual boundaries of the pathosystem and belong to the wider concept of the ecosystem.
Dissemination efficiency	The ability of a parasite to disperse over wide areas in a short time. Dissemination efficiency is a component of the flexibility of the parasite population dynamics and, hence, a factor determining the value of vertical resistance in agriculture.
Diurnal cycle	The cycle resulting from the earth's rotation around its own axis. See also: Seasonal cycle.
Domesticated levels of horizontal resistance	Levels of horizontal resistance which are higher than the natural level because of artificial selection.
Domestication	A deterministic control of evolution in which natural, variable survival values contributing to both the quantity and the quality of the harvestable product are increased above their natural levels, at the expense of survival values contributing to competitiveness in a natural ecosystem which are reduced below their natural levels.
Dynamic system	A system in which one or more patterns undergo change.
Ecosystem	An ecological system; its geographical and conceptual boundaries may be defined as convenient.
Effective initial inoculum	The initial inoculum of a vertical pathotype which matches a stated vertical resistance.
Elimination of vertical resistance	During screening for horizontal resistance, vertical resistance (or its effects) must be eliminated; this can be done genetically, histologically, epidemiologically or by the saturation approach.
Entropy	See: Negative entropy
Environment	All factors which are external to a system but which nevertheless influence that system. Thus the environment of an ecosystem includes various solar system cycles, climate, altitude, latitude, soil, etc.
Environment erosion of horizontal resistance	See: Erosion

Epidemic	In modern epidemiology, this word is used exclusively as a noun and it means one, complete, epidemic cycle. Its adjectival use (e.g. epidemic proportions) is meaningless. In this book, the term applies to both pests and pathogens.
Epidemic cycle	The positive growth of a parasite population from minimum to maximum, followed by its negative growth to minimum. Most epidemic cycles are seasonal.
Epidemiological competence	The ability of a parasite to survive and, hence, to cause an epidemic in a particular environment. See also: Frozen vertical resistance.
Epidemiology	The study of epidemics. In this book, the term includes all parasites and thus embraces insect infestations, etc.
Erosion of horizontal resistance	A loss of horizontal resistance. Host erosion is a genuine loss due to genetical changes in the host population. Parasite erosion is an apparent loss and is due to an increase in horizontal parasitic ability in the parasite population. Environment erosion is also an apparent loss and is due to culivation outside the ecological limits of the host. See also: Vertifolia effect.
Esodemic	That part of the epidemic which is due to auto-infection only.
Evolution	The process of growth (i.e. increases in negative entropy) at the highest systems level in the evolutionary system. See also: Micro-, Macro-evolution.
Exodemic	That part of the epidemic which is due to allo-infection only.
Extinction	The converse of reproduction. At the system levels of the individual organism or cell, extinction normally involves two processes. The first is a loss of behaviour, which is called death; the second is a loss of structure which is called decay or decomposition.
Facultative parasite	A parasite which is also able to grow saprophytically.
Fallacy of the undistributed middle	The fallacy which stems from a non-reversible relationship; thus, "all but not only" (all vertical resistance is oligogenic but not only vertical resistance is oligogenic) or "some but not all" (some horizontal resistance is polygenic but not all horizontal resistance is polygenic). The fallacy leads to false syllogisms, e.g. "all vertical resistance is oligogenic; this resistance is oligogenic; therefore this resistance is vertical".
Feedback	The phenomenon of reciprocation in systems control. Thus, in a thermostatically controlled system, the message to the thermostat that communicates that the system is either too hot or too cold, is called feedback.
Field resistance	This term has no function in a pathosystem context; it is a descriptive term which means the converse of laboratory or glasshouse resistance. However, it is often used, quite incorrectly, to mean horizontal resistance.

Flexibility of population dynamics
The flexibility of the host population dynamics determines the spatial and sequential discontinuity of host tissue under agricultural conditions. The flexibility of the parasite population dynamics determines the ability of the parasite to overcome that discontinuity. Vertical resistance is valuable in agriculture when the host flexibility is maximal and the parasite flexibility is minimal.

Forma specialis
(Plural: *formae speciales;* abbr. f.sp. and f.spp.). A pathotype which is taxonomically above the ranks of polyphyletic pathotype. There are three taxonomic categories of host differentials. If the various differentials are different species, they differentiate *formae speciales* of one parasite species. If they are different inter-specific hybrids, they differentiate polyphyletic pathotypes. If they are all members of one species, they differentiate vertical pathotypes of one parasite species.

Frozen vertical resistance
A vertical genome whose strength is such that the matching vertical pathotype lacks epidemiological competence. This is due to a Hegelian change in vertical genome strength.

Gene-for-gene relationship
A gene-for-gene relationship exists when the presence of a gene in one population is contingent on the continued presence of a gene in another population, and where the interaction between the two genes leads to a single phenotypic expression by which the presence or absence of the relative gene in either organism may be recognised" (Person *et al.*, 1952). That is, a gene for resistance in the host is matched by a gene for parasitic ability in the parasite, and the presence of both is necessary for the demonstration of either. A gene-for-gene relationship invariably confers vertical resistance and vertical parasitic ability. It seems also that all vertical resistance and vertical parasitic ability are invariably the result of a gene-for-gene relationship. A gene-for-gene relationship is demonstrated by the Person, Habgood differential interaction. See also: Second gene-for-gene hypothesis.

Genetic vulnerability
This somewhat meaningless term is often used in place of crop vulnerability with the special implication that the vulnerability is enhanced by crop uniformity.

Geological time
The evolutionary system is a multi-rate system involving many different time scales. Its overall history is measured in the geological time scale which has units of 10^6 years.

Growth
Any increase in negative entropy, which may occur at any systems level in a dynamic system. See also: Reproduction.

Habgood nomenclature
The nomenclature devised by Habgood (1970) to label a series of differential vertical pathodemes or vertical genes. The series of labels is 2^0, 2^1, 2^2, 2^3, etc., each label having an arithmetical value which is double that of its predecessor; thus, 1, 2, 4, 8, 16, etc. The sum of any combinations of labels is unique within the system; thus a vertical genome possessing the first five genes of a series has the sum of 31 and no other combination has this sum. Matching vertical

pathotypes and pathodemes are given the same name which also indicates the composition of the vertical genome. See also: Person/Habgood differential interaction.

Half-life of a vertical pathotype — The disappearence of a vertical pathotype due to negative selection pressure is a logarithmic decrease, mathematically similar to the loss of radio-activity. The strength of vertical genes and genomes can be measured in terms of the relative half-life of the matching vertical pathotypes.

Hart phenomenon — The phenomenon in which a high proportion of sclerenchyma and a thick cuticle restrict the production and liberation respectively of uredospores of *P. graminis*. These mechanisms confer horizontal resistance.

Hegelian change — The qualitative change which results from a small quantitative change. Thus, with small quantitative changes, a substance exhibits qualitative changes of state (liquid, solid, vapour). There have been many Hegelian changes in the course of evolution and cultural development. A special case is Frozen vertical resistance.

Heteroecious — A heteroecious parasite is one whose life cycle can only be completed on two different host species.

Hierarchy — A ranking in which each rank is qualitatively inferior but quantitatively superior to the rank above it.

History — See: Systems history.

Holistic — This word means entire, whole. The holistic approach emphasises the entire system rather than its components.

Homeostasis — The ability of a dynamic system both to maintain a stable optimum and to return to that optimum in the event of a loss of systems balance.

Horizontal — This non-descriptive, abstract term is derived from van der Plank's (1963) classic diagram shown in Figure 2. It labels a category of interaction between pathotypes and pathodemes in which there is constant ranking. Being an abstract term, "horizontal" can be used in many different pathosystem contexts. It can also be re-defined in each context, although each new definition must obviously make sense and must fit the facts. The term can be used to qualify various components of a pathosystem, as listed below.

Horizontal pathodeme — A population of a host in which all individuals have an identical, stated horizontal resistance. One horizontal pathodeme may consist of many different cultivars or botanical varieties which differ in many respects other than their horizontal resistance.

Horizontal pathogenicity, parasitic ability — When a series of different pathodemes of one host species is inoculated with a series of different pathotypes of one parasite species and there is no differential interaction, there can only be a constant ranking. The ranking of the pathotypes, according to pathogenicity or parasitic ability, is constant, regardless of which pathodeme they are tested against. Their pathogenicity (or parasitic ability) is then described as horizontal and they are called horizontal pathotypes. Horizontal pathogenicity (or parasitic ability) is independent of horizontal resistance.

Horizontal pathosystem	The pathosystem sub-system which involves only the interaction of horizontal resistance with horizontal pathogenicity or parasitic ability.
Horizontal pathosystem analysis	The study of both the structure and the behaviour of the horizontal pathosystem.
Horizontal pathosystem management	The management of the horizontal pathosystem with particular emphasis on the system as a whole
Horizontal pathotype	A population of a parasite in which all individuals have an identical, stated, horizontal pathogenicity or parasitic ability.
Horizontal protection/ protective chemical	If a protective chemical is beyond the capacity for micro-evolutionary change of a parasite (e.g. Bordeaux mixture, sulphur, pyrethrins), it is described as horizontal and it confers a horizontal protection.
Horizontal resistance	When a series of different pathodemes of one host species is inoculated with a series of different pathotypes of one parasite species, and there is no differential interaction, there can only be a constant ranking. The ranking of the pathodemes, according to resistance, is constant, regardless of which pathotype they are tested against. Their resistance is described as horizontal and they are called horizontal pathodemes. Horizontal resistance is the only resistance which can reduce the effects of infection after it has occurred; it is the only resistance which can reduce the esodemic. It is consequently universal; it occurs in all plants against all parasites but it is currently inadequate in many cultivars. It is also permanent resistance and it permits cumulative plant breeding. At the histological level of the pathosystem, horizontal resistance is conferred by mechanisms which are beyond the capacity for micro-evolutionary change of the parasite. Such mechanisms are usually numerous, complex and obscure; they may be passive or active mechanisms. They include such phenomena as tolerance and disease-escape which are not strictly resistance mechanisms. Horizontal resistance is independent of horizontal pathogenicity or parasitic ability. Horizontal resistance is usually inherited polygenically; its mechanisms and effects are then quantitative. However, it is occasionally inherited oligogenically and its mechanisms and effects are then qualitative.
Host	A living organism which provides nutrients for a parasite but which is not normally destroyed by that parasite in the way that a prey is destroyed by a predator. The term "host" can be used in the context of an individual, a population or a taxonomic rank. In pathosystem analysis, the deme system is reserved for the host.
Host cycle	The positive growth of the host population from minimum to maximum followed by its negative growth to minimum. The host cycle may refer to entire plants (e.g. annual grasses) or to the components of plants (e.g. the leaves of a deciduous tree).
Host differential	See: Differential hosts.

Host erosion of horizontal resistance	See: Erosion
Hunting	In systems control, hunting is an acceptable fluctuation each side of the optimum. See also: Person model.
Hybrid	The progeny of two parents which are genetically dissimilar. Hence interspecific hybrid, etc. See also: Polyphyletic.
Hybrid seed	Seed produced by the crossing of two inbred lines with a view to increasing hybrid vigour (heterosis). Hybrid seed is used most commonly in maize.
Hybrid spectrum	A spectrum of all degrees of hybridisation between the extremes of the two original parents. See also: Polyphyletic.
Hypersensitivity	An active resistance mechanism in which the rapid death of host cells around the point of infection prevents colonisation. Not all hypersensitivity confers vertical resistance and not all vertical resistance is due to hypersensitivity.
Hypothesis of constant sum of survival values	The hypothesis which states that the sum of variable survival values is constant; one survival value can only be increased at the expense of one or more other survival values.
Immunity	An immune plant is a non-host. Immunity is an absolute quality; it is a non-variable survival value. A plant is either immune or it is not immune to a parasite. Anything less than immunity is resistance, which is a variable survival value. Absolute horizontal resistance differs from immunity in that it can be reduced by host erosion. Vertical resistance differs from immunity in that it cannot reduce the esodemic. Like the word "unique", immunity should be qualified with caution. It can tolerate: absolutely, almost, apparent, nearly, perhaps, quite, really, and surely. It cannot tolerate: comparatively, more, most, rather, somewhat, and very. See also: Polyphyletic resistance
Incomplete vertical resistance	Vertical resistance which confers incomplete protection against non-matching vertical pathotypes. See also: Complete.
Independent system	A system in which the control is fully autonomous.
Individual	The unit of a pattern. In biological systems, the pattern is often called a population and the unit of that pattern is then an individual: thus, individual crop, individual plant, individual cell, etc.
Industrial melanism	The phenomenon in which moth protective colouring changes from white to black with increasing industrial pollution. With reduced pollution the reverse change occurs. These two survival values are mutually exclusive. See also: Hypothesis of constant sum of survival values.
Infection	At the epidemiological level of the pathosystem, infection refers to the contact made between host and parasite: hence, auto-infection, allo-infection. At the histological level, infection refers to the process of penetration of a host by a pathogen.

	The term should not be used comparably with "infestation" (e.g. a severe infection). Nor should it be used as a substitute for "diseased" (e.g. an infected plant). Nor should it be applied to non-hosts (e.g. an infected soil).
Infectious	An infectious disease is due to a pathogen which penetrates its host by infection. A non-infectious disease is not due to parasitism and, in this book, is called a disorder.
Infestation	An epidemic due to a pest (which has mouth parts) as opposed to a pathogen (which lacks mouth parts). In plant pathosystem analysis and management, it is legitimate to use the term epidemic (and its derivatives, esodemic, exodemic) for both pests and pathogens.
Initial inoculum	The total parasite population at the beginning of epidemic. See also: Effective.
Interaction	The effects of a parasite on a host and *vice versa;* the results of such interaction are called disease or damage. See also: Differential.
Juvenile resistance	See: Seedling resistance.
Landrace	The agricultural equivalent of an ecotype
Macro-evolution	Macro-evolution is Darwinian evolution. It involves population changes due to selection pressure. These changes involve characters which are new; they are not reversible (at least in an autonomous system); and they normally occur over periods of geological time. See also: Micro-evolution.
Male gametocide	Any chemical which kills male gametes without damaging female gametes, thus rendering a plant male-sterile but female-fertile.
Manifestation of vulnerability	Crop vulnerability is manifested when the absent parasite or pathotype appears in the pathosystem and the hidden susceptibility to it is revealed.
Markov chain	A multi-decision process in which every decision is entirely random and is independent of any previous decision.
Matching gene theory	See: Gene-for-gene relationship
Matching vertical genes/genomes, etc.	When vertical resistance fails to operate, the vertical genes of the host and the parasite are described as matching. Hence, matching vertical genomes, pathotypes and pathodemes, allo-infection, etc. See also: Over-lapping.
Mathematics of extremes	Mathematical theory associated with variation between a minimum and a maximum.
Mature plant resistance	See: Adult
Maximum horizontal resistance	A point on a scale of measurement of horizontal resistance. It can be determined experimentally and may be a lower level than absolute horizontal resistance.
Micro-evolution	Unlike macro-evolution (Darwinian evolution) micro-evolution involves changes which are not new, which are reversible, and which occur during periods of historical time (i.e. months, years; the foreseeable agricultural future).

Mode/Modal	The largest group of any category of individuals in a mixed population. In a symmetrical biometric curve, the mode is half-way between the minimum and the maximum, and is the same as the mean.
Model	A representation of an actuality. Model building is often called simulation and it is a valid form of speculation. Both ideas and models can be formulated and subsequently tested.
Monoculture	The continuous cultivation of either one crop species or one cultivar in both time and space.
Monocyclic disease	A disease in which the disease cycle coincides with the epidemic cycle.
Multi-decision process	A control process involving many different decisions.
Multi-disciplinary system	A system involving many scientific disciplines.
Multi-level system	A system of many patterns of patterns, each pattern constituting a level.
Multiline	A multiline is a mixture of phenotypically similar but genotypically dissimilar pure lines. The genotypic differences usually involve vertical resistance. A multi line is a plant pattern of vertical resistances.
Multiline effect	An apparent horizontal resistance due to a reduction of the plant exodemic within a multiline.
Multi-rate system	A dynamic system which involves many different time scales. In the evolutionary system, these range from geological time to the movements of atomic particles.
Multi-variate system	A dynamic system with many variables. In the evolutionary system, inherited variables are called survival values.
Natural	Pertaining to nature. A natural system has a control which is fully autonomous. Hence, natural ecosystem, natural pathosystem, natural level of horizontal resistance, etc. See also: Artificial, Crop pathosystem.
Natural level of disease/damage	The amount of disease or parasite damage which normally occurs in a natural pathosystem. This level can be determined in any pathosystem and is the balance which results from the unhampered effects of selection pressures for both horizontal resistance and horizontal pathogenicity (or parasitic ability) on genetically flexible populations of both the host and the parasite, as occurred with African maizes exposed to *Puccinea polysora*.
Natural pathosystem	A pathosystem in a natural, wild ecosystem, in which there has been no interference by man. See also: Crop pathosystem.
Negative entropy	In systems terminology, an improbability of pattern.
Negative-positive rule	The rule that, in screening a flexible host population, a negative screening should be conducted before flowering and a positive screening before, during and/or after harvest.
Negative screening	See: Screening.
Negative selection pressure	See: Selection pressure

Non-matching vertical genes/genomes	If a vertical pathodeme possesses more vertical genes or different vertical genes than/from a vertical pathotype, the vertical genomes do not match and the vertical resistance is effective. See also: Matching.
Non-variable survival values	Inherited characters which are not variable (e.g. immunity).
Obligate parasite	A parasite which is unable to grow saprophytically under natural conditions.
Oligocyclic disease	A disease in which there are only a few disease cycles in each epidemic cycle.
Oligogenic resistance	Resistance whose inheritance is controlled by a few genes. Vertical resistance is always oligogenic resistance but not all oligogenic resistance is vertical.
Open system	A system in which negative entropy can increase when energy is gained from the external environment.
Overlapping vertical genome	The vertical genome of a parasite is described as overlapping when it has more than enough vertical genes to match the vertical genome of a host.
Overwintering	Survival during winter. See also: Aestivation.
Parasite	An organism which depends on a living host for its nutrients but which does not normally destroy that host in the way that a predator destroys its prey. Parasites are usually classified either as pathogens (which lack mouth parts), or pests (which possess mouth parts). The term "parasite" can be used in the context of an individual, a population or a taxonomic rank. In pathosystem analysis, the type-system is reserved for the parasite.
Parasite erosion of horizontal resistance	See: Erosion.
Parasitic ability	The ability of a parasite to survive at the expense of its host. Differences in parasitic ability may be vertical or horizontal. See also: Pathogenicity.
Parasitism	The definitive characteristic of a pathosystem. The term can be used at various systems levels and it describes the phenomenon of a parasite surviving at the expense of its host.
Passive resistance mechanism	A resistance mechanism which is due to innate qualities in the host prior to the attack (e.g. a thick cuticle). It is probable that all passive mechanisms confer horizontal resistance but that not all horizontal resistance is due to passive mechanisms. See also: Active.
Pathodeme	A population of a host species in which all individuals have a stated pathosystem character (resistance) in common. Hence vertical pathodeme, horizontal pathodeme, polyphyletic pathodeme.
Pathogen	A parasite which has no mouth parts and which is studied by a plant pathologist (i.e. fungi, bacteria, viruses). Used in opposition to "pest" which is usually defined as a parasite eating the host by means of mouth parts. The term "pathogen" can be used in the context of an individual, a population or a taxonomic rank.

Pathogen erosion of horizontal resistance See: Erosion.

Pathogenicity The ability of a pathogen to survive at the expense of its host. Differences in pathogenicity may be vertical or horizontal.

Pathosystem Any sub-system of an ecosystem which involves parasitism. A pathosystem may be natural (wild pathosystem, disease triangle) or artificial (crop pathosystem or disease square). The term may be used at any systems level which should be stated. The conceptual, geographical or other boundaries of a pathosystem may be defined as convenient.

Pathosystem analysis The study of pathosystem structure and pathosystem behaviour at all systems levels.

Pathosystem management The deterministic control of a crop pathosystem in order to reduce crop loss due to parasites to a minimum, without reducing yield or quality, without affecting crop uniformity in space or time, without cost to the farmer, and without crop vulnerability.

Pathosystem manager A multi-disciplinary plant scientist who is primarily interested in effective pathosystem management.

Pathotype A population of a parasite species in which all individuals have a stated pathosystem character (pathogenicity or parasitic ability) in common. Hence, vertical pathotype, horizontal pathotype, polyphyletic pathotype.

Pattern An arrangement of units in which the arrangment is usually more important than the units. A pattern has the two intrinsic properties of structure and behaviour which correspond to the spelling and the meaning of a word. The usefulness of vertical resistance in agriculture can be increased by the use of controlled spatial and sequential patterns of plants and crops.

Pedigree breeding A system of plant breeding in which pedigrees are kept of controlled, artificial crosses; the system emphasises genetical studies and a relatively few inherited characters which are studied in detail. It is the system least likely to accumulate horizontal resistance. See also: Population breeding.

Person differential interaction The differential interaction defined by Person (1959) and which can be used to demonstrate a gene-for-gene relationship. See Figure 7.

Person/Habgood differential interaction A Person differential interaction in which the sequence of vertical genomes is that of the Habgood nomenclature. See Figure 8.

Person model A mathematical model designed to demonstrate the permanence of horizontal resistance in a pathosystem in which both the host and the parasite populations are genetically flexible. See Figure 11.

Pest A plant parasite with mouth parts. See also: Pathogen.

Pesticide Any chemical which kills crop parasites or weeds.

Physiologic An obsolete usage meaning pathological or parasitological; hence physiologic race, specialisation, etc.

Physiologic race	An obsolete term meaning vertical pathotype. This terminology cannot distinguish between vertical pathotype and polyphyletic pathotype; or provide a related term to describe a horizontal pathotype. Nor can it provide corresponding terms for host populations.
Physiologic specialisation	An obsolete term meaning any differential interaction between pathodemes and pathotypes. It describes but cannot distinguish between the vertical pathosystem, the polyphyletic pathosystem and the differentiation of *formae speciales*.
Phytosanitation	This term is used solely in the sense of the international, national, regional and local control of plant import and export, for the purposes of preventing the spread of crop parasites.
Plant esodemic	The esodemic at the pathosystem level of one host individual. The plant esodemic involves auto-infection within the components (leaves, etc.) of that one individual. See also: Crop esodemic.
Plant exodemic	The exodemic at the pathosystem level of host individuals. The plant exodemic involves allo-infection between individuals. See also: Crop exodemic.
Plant pattern of vertical resistances	The pattern of different vertical resistances in which each unit of a pattern is an individual plant. Plant patterns are usually spatial only. See also: Crop pattern.
Point	The degree of quantitative change at which a qualitative (Hegelian) change occurs. Thus, boiling point, freezing point, etc.
Polycyclic disease	A disease in which there are many disease cycles in each epidemic cycle.
Polygenic resistance	Resistance whose inheritance is controlled polygenically. All polygenic resistance is horizontal resistance but not all horizontal resistance is polygenic resistance.
Polyphyletic	Polyphyletic means multiple origin. A polyphyletic host or parasite is derived by hybridisation.
Polyphyletic differential interaction	The differential interaction which is observed when a series of different polyphyletic pathodemes is inoculated with a series of different polyphyletic pathotypes (Fig. 14; 15). This differential interaction is a super-system of horizontal pathosystems and is not due to vertical resistance.
Polyphyletic parasitic ability/pathogenicity	Any apparent non-parasitism (or non-pathogenicity) which, however, can be genetically diluted by hybridisation of the parasite.
Polyphyletic pathodemes/pathosystem/pathotypes	The pathosystem, and its pathotypes and pathodemes, which are defined by a polyphyletic differential interaction.
Polyphyletic resistance	Any apparent immunity which, however, can be genetically diluted by hybridisation of the host.
Polyploid	An individual which is derived from one or more progenitors and in which there has been an increase (usually a doubling) of chromosome number.
Population	In biological systems, the pattern is often called a population, and the unit of the pattern is an individual; hence a crop is a population of individual plants. The term "population" can be used in this sense at any systems level.

Population breeding — A system of plant breeding which emphasises a genetically flexible population and the effects of selection pressures on it. See also: Pedigree breeding.

Population dynamics — In a pathosystem sense, population dynamics involve the normal changes which occur in both host and parasite populations in the course of an epidemic and, particularly, those associated with growth, reproduction and dissemination. See also: Flexibility, Micro-evolution.

Population explosion — The exponential growth of a population.

Positive selection pressure — See: Selection pressure.

Positive screening — See: Screening.

Progenitor — The wild species from which a crop species has been domesticated.

Propagule — Any individual of a parasite population which serves the function of dissemination.

Protection — The defence conferred by protective chemicals, as opposed to resistance. Protection (and protective chemicals) may be vertical or horizontal.

Pure line — A homozygous population of a host in which all individuals have been produced by self-pollination and are genetically identical with respect to vertical resistance. See also: Vegetative.

Qualitative resistance — Horizontal resistance is qualitative in its inheritance, its mechanisms and its effects when it is inherited oligogenically. Vertical resistance is qualitative when it confers complete protection against non-matching vertical pathotypes. A polyphyletic pathosystem may also be qualitative.

Quantitative resistance — Horizontal resistance is quantitative in its inheritance, its mechanisms and its effects when it is inherited polygenically. Vertical resistance is quantitative when it confers incomplete protection against non-matching vertical pathotypes. The two kinds of quantitative resistance can be distinguished genetically and histologically. A polyphyletic pathosystem may also be quantitative.

Quarantine — The holding and treating of plants to ensure their freedom from parasites. An effective quarantine station has four attributes which are discussed at p. 136.

R-gene — A vertical resistance gene. R-gene is dominant and r-gene is recessive.

Race — When used as a term, this word is one of the least precise in the whole of biology. It means any sub-division of any species (host or parasite) defined by any criteria (morphology, physiology, pathology, etc.). Thus, it can be a vertical, horizontal or polyphyletic pathotype or pathodeme; *forma specialis;* etc. See also: Physiologic race.
When used as a name, the word "race" is acceptable. Thus Race 1 of *Phytophthora infestans*.

Rarity of contact — Vertical resistance reduces the intensity of the exodemic by increasing the rarity of successful (matching) contact between host and parasite. In agriculture, rarity of contact is achieved by increasing the rarity of vertical pathodemes by the use of patterns of host populations, and by increasing the rarity of vertical pathotypes by the use of strong vertical genes.

Ratoon — A crop which is produced by allowing a harvested crop to grow again. Ratoon crops are typical of sugarcane but are also used occasionally with other crop species.

Recovery of vertical resistance — The converse of a breakdown. Thus, with a leaf disease of a deciduous tree, the esodemic ends with leaf-fall. With refoliation, the vertical resistance is again effective against all non-matching allo-infection and, because the esodemic cannot begin until there is a matching allo-infection, the vertical resistance is said to have "recovered".

Relative half-life — See: Half-life.

Repetitive plant breeding — The breeding required to produce a sequence of closely similar cultivars in a boom-and-bust cycle due to vertical pathosystem mismanagement. See also: Cumulative plant breeding.

Reproduction — A special form of growth in which an entire pattern is reproduced. Reproduction may occur at any systems level.

Reproductive capacity — Some parasites have a higher capacity for reproduction than others. Vertical resistance is likely to be more valuable against parasites with a low reproductive capacity.

Resistance — At the epidemiological level of the pathosystem, resistance is the ability of the host population to reduce the epidemic or infestation. Vertical resistance can only reduce the exodemic and the esodemic can only be reduced by horizontal resistance. At a lower systems level, one host individual possesses many different resistances to many different parasites. At a still lower systems level, each resistance is conferred by one or more resistance mechanisms.
Resistance may also occur against disorders, e.g. frost resistance.

Resistance mechanism — At the pathosystem level of the individual host, there are many different mechanisms of resistance to any one parasite. Each mechanism confers either vertical or horizontal resistance; there are no other possibilities. Many mechanisms remain unidentified. Those which have been identified are labelled with descriptive terms which should never be employed as alternatives for the abstract terms vertical and horizontal. There are various classifications of resistance mechanisms; thus, active and passive mechanisms, physiological and physical, etc. Vertical resistance is usally conferred by a single, simple, active mechanism; horizontal resistance is usually conferred by many mechanisms, most of which are variable, complex, and unidentified.

Restoration of horizontal resistance — The converse of erosion of horizontal resistance; the accumulation of horizontal resistance in an abnormally susceptible host.

Rossetto hypothesis — The hypothesis of Rossetto (1975) which states that a non-uniform dispersal of a parasite in space is a positive survival value for the parasite, because it reduces selection pressure for resistance in the host.

Saprophyte — A non-parasitic plant or micro-organism which normally lacks chlorophyll and derives its nutrients from dead organic matter.

Screening — A means of exerting selection pressure on a genetically flexible population. Screening may be positive (retention of desirable individuals) or negative (rejection of undesirable individuals). See also: Negative-positive rule.

Seasonal cycle — The cycle resulting from the earth's rotation round the sun.

Second gene-for-gene hypothesis — "In host-parasite systems in which there is a gene-for-gene relationship, the quality of a resistance gene in the host determines the fitness of the matching virulence gene in the parasite to survive when this virulence is unnecessary; and, reciprocally, the fitness of the virulence gene to survive when it is unnecessary determines the quality of the matching resistance gene as judged by the protection it can give the host". (van der Plank, 1975).

Seedling resistance — Resistance which is apparent in seedlings. This phenomenon can be used to distinguish vertical resistance, which operates in seedlings, from horizontal resistance which is often, but not necessarily, more marked in adult plants.

Selection pressure — Any factor which leads to a change in one or more variable survival values. Selection pressure may be positive; it then leads to an increase in the survival values. Or it may be negative; it then leads to a decrease in the survival values. In artificial selection, selection pressure is exerted by the process of screening.

Sequential — See: Discontinuity and Pattern.

Sett — A piece of sugarcane stem used for propagation.

Simple vertical genome — A vertical genome with only a few vertical genes; hence simple vertical pathotype, pathodeme, pathogenicity (or parasitic ability), resistance, etc.

Simulation — A form of speculation by model building, often with a computer.

Source of resistance — Traditionally, plant breeders assumed that it was impossible to breed for pest/pathogen resistance without first finding a good source of resistance. It is now clear that horizontal resistance can be accumulated in susceptible cultivars without any other source of resistance.

Spatial — See: Discontinuity and Pattern.

Species memory — The evolutionary Hegelian change which permits the growth of memory at the species level. The species memory can grow indefinitely because the species endures for geological time, whereas the individual memory endures for only one life-span. The growth of the species memory in man has been exponential and has itself exhibited many Hegelian changes of increasing frequency and magnitude. It is otherwise called cultural development.

Spectrum — In the conceptual sense, a range of all differences in degree between two extremes. Hence, hybrid spectrum; spectrum of domestication, etc.

Static system	A system in which the patterns never change. Thus a book is a static system of chapters, paragraphs, sentences and words; a taxonomic system is also static.
Stochastic	A stochastic decision process is one in which any decision is random but is also related, on a probability basis, to the state of the system at the time of the decision.
Stoetzer strategy	A breeding strategy designed to accumulate horizontal resistance. See Figure 13.
Strong protective chemical	A vertical protective chemical whose matching pathotype is both rare and becomes rare quickly after having been common.
Strong vertical gene/genome	A vertical gene or genome whose matching vertical pathotype is both rare and becomes rare quickly after having been common.
Structure	Systems structure corresponds to the spelling of a word, as opposed to systems behaviour which corresponds to the meaning of a word.
Subsistence cultivar/crop	A crop or cultivar cultivated by a peasant farmer for the supply of himself and his family, c.f. cash crop which is cultivated for sale and income. Subsistence cultivars are often relatively primitive in terms of artificial selection; they have relative low yields and quality but high levels of horizontal resistance.
Sub-system	A system within a larger system.
Super race	The complex vertical pathotype which matches all the simple vertical pathodemes of a multiline.
Super-system	A system of sub-systems.
Survival value	Any inherited character which contributes to evolutionary survival. Survival values may be quantitatively variable (e.g. most horizontal resistance), qualitatively variable (e.g. most vertical resistance) or non-variable (e.g. immunity).
Susceptibility	Susceptibility is the opposite of resistance and is inversely proportional to it. Susceptibility may be vertical or horizontal.
Symmetrical system	A system in which the same pattern is repeated at every system level (See Fig. 8).
Synergy	The phenomenon in which the total effect is greater than the sum of the parts.
System	A pattern of patterns or, more commonly, many patterns of patterns which are inter-related so as to constitute one, overall pattern.
Systemic disease	A disease in which the pathogen, usually a virus, invades all the tissues of a host individual.
Systems analysis	The study of systems structure and systems behaviour at all systems levels.
Systems balance	The state of dynamic equilibrium which is normally maintained in a dynamic system. A loss of systems balance leads to instability and, possibly, to a self-destructive system.
Systems behaviour	If we consider a word as the basic example of a pattern, its spelling is the equivalent of systems structure and its meaning is the equivalent of systems behaviour.

Systems concept	The theoretical treatment of systems. The systems concept is normally subdivided into systems analysis and systems management.
Systems control	The communication between sub-systems which preserves systems balance in a dynamic system. The study of systems control is called cybernetics. Systems control may be autonomous or deterministic.
Systems history	Every dynamic system has a history which can be measured in the units of an appropriate time scale. Systems history is analysed mathematically with recurrence formulae which express the state of the system at time $(t+1)$ as a function of the state of the system at time t.
Systems management	The purposeful control of a system in order to achieve predetermined objectives.
Systems structure	If we consider a word as the basic example of a pattern, its spelling is the equivalent of systems structure and its meaning is the equivalent of systems behaviour.
Teleology	The doctrine of deterministic control in a natural system. Teleology was important only for so long as the fully autonomous control of natural systems was in dispute.
Tolerance	When two cultivars are equally diseased, the more tolerant one suffers a smaller loss of yield. See p. 75 for full discussion.
Type-system	A temporary, artificial classification of populations of a parasite; this system is parallel to the deme-system which is reserved for populations of the host. Hence, pathotype, pathodeme.
Universal susceptible	A vertical pathodeme which possesses no vertical genes.
Unselected level of horizontal resistance	The level of horizontal resistance which is reached in a genetically flexible host population in the absence of the parasite. This is a point on a scale of measurement of horizontal resistance and it can be determined experimentally.
V-gene	A vertical pathogenicity (or parasitism) gene. V-gene is dominant and v-gene is recessive. Note that the letter v stands for "virulence" not "vertical". See also: R-gene
Variable survival value	An inherited character which is variable between a minimum and a maximum.
Vegetative propagation/reproduction	Propagation or reproduction without sexual processes which leads to the formation of a clone in which all individuals are genetically identical except for somatic mutation. The individual plants of a clone are the epidemiological equivalent of the individual leaves of one tree. See also: Plant esodemic, Crop esodemic.
Vertical	This non-descriptive, abstract term is derived from van der Plank's (1963) classic diagram shown in Figure 1. It labels a category of interaction between pathotypes and pathodemes called the Person differential interaction. Being an abstract term, "vertical" can be used in many different pathosystem contexts. It can also be re-defined in each

context, although each new definition must obviously make sense and must fit the facts. The term can be used to qualify various components of a pathosystem as listed below.

Vertical gene — One of a pair of matching genes in a gene-for-gene relationship: one gene occurs in the host, and the other occurs in the parasite. But, because they are matching genes, they can be given the same name. In general discussion, it is often convenient to refer to a vertical gene without necessarily indicating whether it is in the host or the parasite.

Vertical genome — All the vertical genes present in a vertical pathotype or pathodeme.

Vertical mutability of the parasite/pathogen — The readiness with which a parasite produces a "new" vertical pathotype, regardless of whether such production is by mutation (in the strict sense), sexual or parasexual means or by a population increase of a previously rare pathotype.

Vertical pathodeme — A population of a host in which all individuals have an identical, stated vertical resistance. One vertical pathodeme may consist of many different cultivars or botanical varieties which differ in many respects other than their vertical resistance.

Vertical pathogenicity/parasitic ability — When a series of different pathodemes of one host species is inoculated with a series of different pathotypes of one parasite species and the levels of parasite damage show a Person differential interaction, the pathogenicity (or parasitic ability) of the pathotypes is described as vertical and they are called vertical pathotypes.

Vertical pathosystem — The pathosystem sub-system which involves only the interaction of vertical resistance with vertical pathogenicity or parasitic ability.

Vertical pathosystem analysis — Analysis of the vertical pathosystem which, in practice, is dependent primarily on the distinction between auto-infection and allo-infection.

Vertical pathosystem management — Management of the vertical pathosystem to ensure that contacts between host and parasite are rare. Rarity of vertical pathodemes is achieved by the use of patterns of hosts, and rarity of vertical pathotypes is achieved by the use of strong vertical genes. Frozen vertical resistance may also be obtainable.

Vertical pathotype — A population of a parasite in which all individuals have an identical, stated, vertical pathogenicity or parasitic ability.

Vertical protection/protective chemical — If a protective chemical is within the capacity for micro-evolutionary change of a parasite (e.g. DDT, benomyl fungicides), it is described as vertical and it confers a vertical protection, which may be strong or weak.

Vertical resistance — When a series of different pathodemes of one host species is inoculated with a series of different pathotypes of one parasite species and the levels of parasite damage show a Person differential interaction (Figure 7), the resistance of the pathodemes is described as vertical and they are called vertical pathodemes. Vertical resistance can only prevent infection. It can only prevent allo-infection and, because some matching always occurs, it cannot prevent all allo-infection. Its sole function is thus to reduce the exodemic. In a crop pathosystem which has spatial and sequential continuity of host tissue, the exodemic occurs at the population level of the system. In such a crop pathosystem, vertical resistance is temporary; it can only delay the onset of the crop esodemic. Its use then leads to a boom-and-bust cycle of cultivar production, and repetitive plant breeding. In a natural pathosystem exhibiting spatial and sequential discontinuity of host tissue, the effects of vertical resistance are permanent and are those of an apparent horizontal resistance.

At the histological level of the pathosystem, vertical resistance is conferred by mechanisms which are within the capacity for micro-evolutionary change of the parasite. Such mechanisms probably always occur singly and they are simple, prominent and active mechanisms. At the biochemical level, each mechanism probably involves a single chemical pathway which, however, may be complex.

At the genetical level, vertical resistance is always inherited oligogenically but not all oligogenic resistance is vertical. It is probable that all vertical resistance involves a gene-for-gene relationship and *vice versa*.

Vertical susceptibility — The converse of vertical resistance. Vertical resistance is due to the presence of one or more vertical genes; vertical susceptibility is due to their absence.

Vertifolia effect — A host erosion of horizontal resistance which occurs during breeding for vertical resistance. The maximum vertifolia effect is the unselected level of horizontal resistance.

Virulence — This ambiguous term is often used to mean vertical pathogenicity but it is also used to mean horizontal pathogenicity.

Vulnerability — See: Crop vulnerability.

Weak protective chemical — A vertical protective chemical against which the matching pathotype is common.

Weak vertical gene/genome — A vertical gene or genome whose matching vertical pathotype is common.

References

Abbott, E. V.: U.S.D.A. Tech. Bull. No. 641, pp. 96 (1938).

Abbott, E. V.: In: Martin, J. P., Abbott, E. V., Hughes, C. G. (Eds.): Sugarcane Diseases of the World, Vol. 1, pp. 542. Amsterdam, London, New York: Elsevier Publ. Co. 1961.

Ackoff, R. L.: Towards a system of system concepts. Management Sci. **17**, 11 (1971).

Bettencourt, A. J., Noronha-Wagner, M.: Genetic Study of the resistance of *Coffea* spp. to leaf rust. I. Identification and behaviour of four factors conditioning disease reaction in *Coffea arabica* to twelve physiologic races of *Hemileia vastatrix*. Can. J. Botany **45**, 2021—2031 (1967).

Bettencourt, A. J., Noronha-Wagner, M.: Genetic factors conditioning resistance of *Coffea arabica* L. to *Hemileia vastatrix*. Berk. and Br. Agronomia Lusit. **31**, 285—292 (1971).

Biffen, R. H.: Mendel's law of inheritance and wheat breeding. J. Agr. Sci. **1**, 4—48 (1905).

Bigornia, A. E.: Pers. comm. Guinobatan, The Philippines (1972).

Black, W., Mastenbroek, C., Mills, W. R., Petersen, L. C.: A proposal for an international nomenclature of races of *Phytophthora infestans* and of genes controlling immunity in *Solanum demissum* derivatives. Euphytica **2**, 173—178 (1953).

Booth, C.: The Genus Fusarium, pp. 237. Kew, England: Commonwealth Mycol. Inst. 1971.

Bruce, J.: Pers. comm. Ministry of Overseas Development London (1973).

Cammack, R. H.: Trans. Brit. Mycol. Soc. **42**, 27—32 (1959).

Cannon, W. B.: The Wisdom of the Body. New York: Norton and Co., Inc. 1939.

Chiarappa, L.: Crop Loss Assessment Methods. England: F.A.O. and Commonwealth Mycol. Inst. 1971.

Chona, B. L., Padwick, G. W.: More light on the red rot epidemic. Indian Farming **3**, 70—73 (1942).

Coons, G. H.: Progress in plant pathology, control of disease by resistant varieties. Phytopathology **27**, 622—632 (1937).

Crosse, J. E.: Plant Pathogenic Bacteria in Soil. In: Gray, T. R. G., Parkinson, D. (Eds.). The Ecology of Soil Bacteria, an International Symposium, 552—572. Liverpool: Univ. Press 1968.

Day, P. R.: Genetics of Host-Parasite Interaction, pp. 238. San Francisco: W. H. Freeman and Co. 1974.

Dickinson, P. J.: Pers. common.. London: Ministry of Overseas Development 1972.

Ferwerda, F. P.: Rubber. In: Ferwerda, F. P., Witt, F. (Eds.): Outlines of Perennial Crop Breeding in the Tropics, Misc. Paper, Vol. **4**, pp. 427—458. Wageningen, Netherlands: Landbouwhogeschool 1969.

Flor, H. H.: Inheritance of pathogenicity in *Melampsora lini*. Phytopathology **32**, 653—669 (1942).

Flor, H. H.: Genetic Controls of Host Parasite Interactions in Rust Diseases. In: Holton, C. S., Fischer, G. W., Fulton, R. W., Hart, H., McCallan, S. E. A. (Eds.): Plant Pathology, Problems and Progress, 1908—1958, pp. 137—144. Madison: Univ. Wis: Press 1959.

Gibson, R. W.: Glandular hairs providing resistance to aphids in certain wild potato species. Ann. Appl. Biol. **68**, 113—119 (1971).

Gilbert, L. E.: Butterfly-plant coevolution: Has *Passiflora adenopoda* won the selectional race with Heliconiine butterflies? Science **172**, 585—586 (1971).

Gilmour, J. S. L., Heslop-Harrison, J.: The deme terminology and the units of micro-evolutionary change. Genetica **27**, 147—161 (1954).

Graaff, N. A., van der: Pers. comm. Inst. Agr. Res. Jimma, Ethiopia (1975).

Green,G.J.: Physiologic races of wheat stem rust in Canada from 1919 to 1969. Can. J. Botany **49**, 1575—1588 (1971).
Habgood,R.M.: Designation of physiological races of plant pathogens. Nature **227**, 1268—1269 (1970).
Hart,H.: Morphologic and physiologic studies on stem rust resistance in cereals. U.S.D.A. Tech. Bull. **266**, p. 76 (1939).
Hocker,A.L., le Roux,P.M.: Sources of protoplasmic resistance to *Puccinia sorghi* in corn. Phytopathology **47**, 187—191 (1957).
Hutchinson,F.W.: Defoliation of *Hevea brasiliensis* by aerial spraying. J. Rubber Res. Inst., Malaya **15**, 241 (1958).
James,W.C.: An illustrated series of assessment keys for plant diseases, their preparation and usage. Can. Plant Dis. Survey **51**, 39—65 (1971).
James,W.C.: A manual of Assessment Keys for Plant Diseases. Publication No. 1458 Agriculture, Canada (1973).
James,W.C., Shih,C.S., Callbeck,L.C., Hodgson,W.A.: Interplot interference in field experiments with late blight of potato *(Phytophtora infestans)*. Phytopathology **63**, 1269—1275 (1973).
Jensen,N.F.: Intra-varietal diversification in oat breeding. Agron,J. **44**, 30—34 (1952).
Johnsson,R., Stubbs,R.W., Fuchs,E., Chamberlain,N.H.: Nomenclature for physiologic races of *Puccinia striiformis* infecting wheat. Trans. Brit. Mycol. Soc. **58**, 475—480 (1972).
Keiding,J.: Persistance of resistant populations after the relaxation of the selection pressure. World Rev. Pest Contr. **6**, 115—130 (1967).
Kommedahl,T., Christensen,J.J., Frederiksen,R.A.: A half century of research in Minnesota on flax wilt caused by *Fusarium oxysporum*, pp. 35. Tech. Bull. No. **273**, Agr. Expt. Stn., Univ. of Minnesota 1970.
Kranz,J.: Epidemics of Plant Diseases. Mathematical Analysis and Modeling, pp. 170. Springer: Berlin-Heidelberg-New York 1974a.
Kranz,J.: Epidemiology, Concepts and Scope. In: Raychaudhuri,S.P., Verma,J.P. (Eds.): Current Trends in Plant Pathology, pp. 26—32. Lucknow Univ. Dept. Bot. 1974b.
Lerner,I.M.: The Genetic Basis of Selection, pp. 298. New York: Wiley and Sons 1958.
Link,D., Rossetto,C.J.: Rev. Per. Entomol. **15**, 225—227 (1972).
Lupton,F.G.H., Johnson,R.: Breeding for mature plant resistance to yellow rust in wheat. Ann. Appl. Biol. **66**, 137—143 (1970).
Martin,J.P., Han Lioe Hong, Wismer,C.A.: In: Sugarcane Diseases of the World, Vol. I, pp. 542. Amsterdam-London-New York: Elsevier Publ. Co. 1961.
Mayne,W.W.: Physiologic specialisation in *Hemileia vastatrix*. Berk. and Br. Nature **129**, 510 (1932).
Mayne,W.W.: Ann. Rep. of the Coffee Sci. Offiver, 1934—1935, pp. 28. Dept. of Agr., Mysore State 1935.
Mayne,W.W.: Ann. Rep. of the Coffee Sci. Officer, 1935—1936, pp. 21. Dept. of Agr., Mysore State 1936.
McDonald,J.: A preliminary account of a disease of green coffee berries in Kenya colony. Trans. Brit. Mycol. Soc. **11**, 145—154 (1926).
Medawar,P.B.: The Art of the Soluble, pp. 160. London: Methuen 1967.
Miller,J.I.: The Spice Trade of the Roman Empire, pp. 294. Oxford: Clarendon Press 1969.
Müller,K.O.: Affinity and reactivity of angiosperms to *Phytophtora infestans*. Nature **166**, 392—394 (1950).
Nattrass,R.M.: E. Afr. Agr. J. **18**, 39 (1952).
Nutman,F.J., Roberts,F.M.: Studies on the biology of *Hemileia vastatrix* Berk. and Br. Trans. Brit. Mycol. Soc. **46**, 27—48 (1963).
Painter,R.H.: Insect Resistance in Crop Plants, pp. 520. New York: MacMillan 1951.
Patten,B.C.: Systems Analysis and Simulation in Ecology, Vol. I (1971), Vol. II, pp. 607, 592. New York-London: Academic Press 1971.

Person, C. O.: Gene-for-gene relationships in host parasite systems. Can. J. Botany **37**, 1101—1130 (1959).

Person, C. O.: Pers. comm. Vanvouver: Univ. Brit. Columbia 1975.

Person, C. O., Samborski, D. J., Rohringer, R.: The gene-for-gene concept. Nature **194**, 561—562 (1952).

Person, C. O., Sidhu, G.: Proceedings of a Panel on Mutation Breeding for Disease Resistence, pp. 31—38. Vienna: I.A.E.A. 1971.

Person, C. O., Sidhu, G.: Pers. comm. Vancouver: Univ. Brit. Columbia 1972.

Plank, J. E., van der: Plant Diseases. Epidemics and Control, pp. 349. New York-London: Academic Press 1963.

Plank, J. E., van der: Disease Resistance in Plants, pp. 206. New York-London: Academic Press 1968.

Plank, J. E., van der: Stability of resistance to *Phytophtora infestans* in cultivars without R-genes. Potato Res. **19**, 263—270 (1971).

Plank, J. E., van der: Principles of Plant Infection, pp. 216. New York-London: Academic Press 1975.

Punithalingam, E., Gibson, I. A. S.: CMI Descriptions of pathogenic fungi and bacteria, No. 330. Commonwealth Mycol. Inst., England (1972).

Robinson, H. J.: Renascent Rationalism. Toronto: MacMillan of Canada 1975.

Robinson, R. A.: Disease resistance terminology. Rev. Appl. Myol. **48**, 593—606 (1969).

Robinson, R. A.: Vertical resistance. Rev. Plant Path. **50**, 233—239 (1971).

Robinson, R. A.: Horizontal resistance. Rev. Plant P. **52**, 483—501 (1973a).

Robinson, R. A.: The search and need for horizontal resistance to coffee rust and prospects for similar resistance to CBD in Ethiopia. Proc. Consulta Conjunta de Expertos Sobre La Prevencion de la Roya Cafeto, pp. 181. Turrialba, Costa Rica: FAO/IICA, OEA 1973b.

Robinson, R. A.: Potato development. UNDP Rep. TA 3208 to the Gov. Kenya pp. 33. Rome: FAO/UN 1973c.

Robinson, R. A.: Terminal report of the FAO coffee pathologist to the goverment of Ethiopia, pp. 17. Rome: FAO AGO/74/443 1974.

Rossetto, C. J.: Pers. comm. Inst. Agr. Campinas, S. P., Brazil 1975.

Rowell, P. L., Miller, D. G.: Induction of male sterility in wheat with 2-chloroethylphosphonic acid (Ethrel). Crop Sci. **11**, 629—631 (1971).

Schrödter, H., Ullrich, J.: Untersuchungen zur Biometeorologie und Epidemiologie vor *Phytophthora infestans* (Mont.) de By. auf mathematisch-statistischer Grundlage. Phytophatology Z. **58**, 87—193 (1965).

Shaw, D. E.: Eradication of coffee rust in Papua in 1965. Proc. Cons. Expert. Prev. Roya del Cafeto, pp. 181. Turrialba, Costa Rica: FAO/IICA, OEA 1973.

Simmonds, N. W.: Studies of the tetraploid potatoes III, progress in the re-creation of the tuberosum group. J. Linn. Soc. (Bot.), **59**, 229—288 (1966).

Simmonds, N. W.: Prospects of Potato Improvement. In: 48th Ann. Rep. Scot. Plant Breeding Stat., pp. 64, 1968—1969. Edinburgh (1969).

Stanton, W. R., Cammack, R. H.: Resistance to the maize rust *Puccinia polysora* Underw. Nature **172**, 505—506 (1953).

Stevens, N. E.: Disease, damage and pollination types in grains. Science **89**, 339—340 (1939).

Stevens, N. E.: Botanical research by unfashionable technics. Science **93**, 172—176 (1941).

Stewart, D. M., Romig, R. W., Rothman, P. G.: Distributions and prevalence of physiologic races of *Puccinia graminis* in the United States in 1968. Plant Dis. Reporter **54**, 256—260 (1970).

Stoetzer, H. I. A.: Pers. common. Awassa Exp. Stat. Awassa, Ethiopia (1975).

Storey, H. H., Ryland, A. K.: Ann. Rep. E. Afr. Agr. For. Res. Org., Nairobi (1955).

Storey, H. H., Howland, A. K., Hemingway, J. S., Jameson, J. D., Baldwin, B. J. T., Thorpe, H.C., Dixon, G.E.: East African work on breeding maize resistant to the tropical American rust *Puccinia polysora*. Empire J. Exp. Agr. **26**, 1—17 (1958).

Waller, J.M.: Sugarcane smut (*Ustilago scitaminea*) in Kenya. I Epidemiology. Trans. Brit. Mycol. Soc. **52**, 139—151 (1969).

Waller, J.M.: Sugarcane smut (*Ustilago scitaminea*) in Kenya. II Infection and resistance. Trans. Brit. Mycol. Soc. **54**, 405—414 (1970).

Watt, K. E. F. (Ed.): Systems Analysis in Ecology, pp. 276. New York-London: Academic Press 1966.

Wiener, N.: The Human Use of Human Beings, pp. 241. London: Eyre and Spottiswoode 1950.

Index